INFORMATION AND THE NATURE OF REALITY

T0177165

INFORMATION AND THE NATURE OF REALITY

FROM PHYSICS TO METAPHYSICS

Edited by

PAUL DAVIES
Arizona State University

NIELS HENRIK GREGERSEN
Copenhagen University

CAMBRIDGE
UNIVERSITY PRESS

CAMBRIDGE
UNIVERSITY PRESS

Shaftesbury Road, Cambridge CB2 8EA, United Kingdom

One Liberty Plaza, 20th Floor, New York, NY 10006, USA

477 Williamstown Road, Port Melbourne, VIC 3207, Australia

314–321, 3rd Floor, Plot 3, Splendor Forum, Jasola District Centre,
New Delhi – 110025, India

103 Penang Road, #05–06/07, Visioncrest Commercial,
Singapore 238467

Cambridge University Press is part of Cambridge University Press & Assessment,
a department of the University of Cambridge.

We share the University's mission to contribute to society through the pursuit of
education, learning and research at the highest international levels of excellence.

www.cambridge.org
Information on this title: www.cambridge.org/9781107684539

First published 2010
Reprinted 2011
Canto Classics edition 2014 (version 2, February 2023)

A catalogue record for this publication is available from the British Library

ISBN 978-1-107-68453-9 Paperback

TO THE MEMORY OF

ARTHUR R. PEACOCKE

(1924–2006)

CONTENTS

vii

Contents

ABOUT THE AUTHORS

PHILIP CLAYTON (PhD Yale, 1986) is Ingraham Professor of Theology, Claremont School of Theology, and Professor of Religion and Philosophy, Claremont Graduate University. His areas of research include science and religion, philosophical theology, constructive theology, and metaphysics. He is author of *God and Contemporary Science* (1998) and *Mind and Emergence* (2004) and he has more recently written *Adventures in the Spirit* (2008) and *In Quest of Freedom: The Emergence of Spirit in the Natural World* (2009). Among his influential edited books are *The Re-Emergence of Emergence* (2006, with Paul Davies) and *The Oxford Handbook of Religion and Science* (2006).

PAUL DAVIES (PhD London, 1970) has held university positions at Cambridge, London, Newcastle, Adelaide, and Sydney before joining Arizona State University as Professor and Director of Beyond: Center for Fundamental Concepts in Science. He has helped develop quantum field theory, and alongside his scientific work in cosmology and atomic astrophysics he has maintained interests in the origin of time asymmetry and astrobiology. He has written more than 25 books, both popular and specialist works, including *About Time* (1995), *The Fifth Miracle* (1998), *How to Build a Time Machine* (2002), *The Goldilocks Enigma* (2007) and *The Eerie Silence* (2010). He has received many awards, including the 2001 Kelvin

Medal, the 2002 Faraday prize, and in 1995 the Templeton Prize.

TERRENCE W. DEACON (PhD Harvard, 1984) taught for many years at Harvard University and Boston University, until in 2002 he joined the Department of Anthropology and the Helen Wills Neuroscience Institute at University of California, Berkeley. In his many articles and book chapters, Deacon's research combines human evolutionary biology and neuroscience in the investigation of the evolution of human cognition. His acclaimed book, *The Symbolic Species: the Co-Evolution of Language and Brain*, was published in 1997, and has been translated into several languages. In 2007 he was awarded the Staley Prize by the School of American Research.

NIELS HENRIK GREGERSEN (PhD Copenhagen University, 1987) is Professor of Systematic Theology at Copenhagen University, and Co-Director of the Centre for Naturalism and Christian Semantics. His areas of research are contemporary theology and science and religion, with a special emphasis on the complexity sciences and current developments in evolutionary biology. He is author of four books and more than 150 scholarly articles. He has edited or co-edited 15 books on science and religion, including *Design and Disorder* (2001), *From Complexity to Life* (2003), and *The Gift of Grace* (2005).

JOHN F. HAUGHT (PhD Catholic University, 1970) is Senior Fellow, Science & Religion, Woodstock Theological Center, Georgetown University. His area of specialization is systematic theology, with a particular interest in issues pertaining to science, cosmology, evolution, ecology, and religion. He is the author of 17 books, most of them on the subject of science and religion. His latest books are: *God and the New Atheism: A Critical Response to Dawkins, Harris and Hitchens* (2008) and *Making Sense of Evolution: Darwin, God, and the Drama of Life* (2010).

JESPER HOFFMEYER is Professor Emeritus in the Department of Molecular Biology, University of Copenhagen. Hoffmeyer did work in experimental biochemistry in the 1970s, but gradually turned his research interests towards questions of theoretical biology: since 1988 in biosemiotics. Recent publications include *A Legacy of Living Systems: Gregory Bateson as Precursor to Biosemiotics* (2008) (as editor) and his major work *Biosemiotics: An Examination into the Signs of Life and the Life of Signs* (2008).

BERND-OLAF KÜPPERS studied physics and mathematics at the universities of Bonn and Göttingen. From 1971 to 1993 he worked at the Max Planck Institute for Biophysical Chemistry in Göttingen. Since 1994 he has been Professor of Natural Philosophy at the Friedrich Schiller University of Jena, and since 2008 also Director of the Frege Centre for Structural Sciences. Books include *Molecular Theory of Evolution* (1985), *Information and the Origin of Life* (1990), *Natur als Organismus* (1992), *Nur Wissen kann Wissen beherrschen* (2008) and *Wissen statt Moral* (2010). He is a member of Germany's National Academy of Sciences and of the Academia Europaea, London.

SETH LLOYD received his BA in physics from Harvard, his MPhil in philosophy of science from Cambridge, and his PhD in physics from Rockefeller University under the supervision of Heinz Pagels. After working at Caltech and Los Alamos, he joined the faculty at MIT, where he is Professor of Quantum-Mechanical Engineering. His research focuses on how physical systems process information. Lloyd was the first person to develop a realizable model for quantum computation. He is the author of *Programming the Universe* (2006) and is currently the Director of the W. M. Keck Center for Extreme Quantum Information Theory at MIT.

ERNAN MCMULLIN is the O'Hara Professor Emeritus of Philosophy and Founder Director of the Program in History and Philosophy of Science at the Notre Dame University. As a philosopher of science he has written and lectured extensively on subjects ranging from the relationship between cosmology and theology to the impact of Darwinism on Western religious thought. Among the books he has written or edited are *The Concept of Matter* (1963), *Newton on Matter and Activity* (1978), *Construction and Constraint: The Shaping of Scientific Rationality* (1988), *The Philosophical Consequences of Quantum Theory* (with James Cushing, 1989), *The Church and Galileo* (2005).

ARTHUR R. PEACOCKE (1924–2006) was a biochemist and theologian from Oxford University. Having taught at Birmingham he returned to Oxford in 1959 as Professor of Physical Biochemistry. In this capacity he published more than 125 papers and three books. Later he resumed his theological interests, became ordained in 1971, and went to serve as Dean of Clare College, Cambridge University. In 1985 he became the founding director of the

Ian Ramsey Centre, at Oxford. In 1992–1993 he gave the Gifford Lectures, published as *Theology for a Scientific Age* (1993). In a series of books, beginning with *Science and the Christian Experiment* (1971) and ending with *All That Is: A Naturalistic Faith for the Twenty-First Century* (2007), he laid the groundwork for a generation of younger scholars in the field of science and religion. In 2001 he was awarded the Templeton Prize.

HOLMES ROLSTON, III is University Distinguished Professor and Professor of Philosophy Emeritus at Colorado State University. He has written eight books, including *Science and Religion: A Critical Survey* (most recent edition 2006), *Environmental Ethics* (1988), and *Three Big Bangs: Matter-Energy, Life, Mind* (2010). He gave the Gifford Lectures, University of Edinburgh, 1997–1998, published as *Genes, Genesis and God* (1999). He has lectured on all seven continents. He was named laureate for the Templeton Prize in Religion in 2003. A recent intellectual biography is *Saving Creation: Nature and Faith in the Life of Holmes Rolston III*, by Christopher J. Preston (2009).

JOHN MAYNARD SMITH (1920–2004) was a geneticist and theoretical evolutionary biologist. In the late 1950s and early 1960s he did pioneering work on the genetics of aging in fruit flies, and wrote *The Theory of Evolution* (1958). As the Founding Dean of the School of Biological Sciences at the University of Sussex (1965–1985), his interests turned into theoretical problems of evolutionary biology, especially concerning the relation between mathematics and life. He formalized the Evolutionary Stable Strategy (EES), today a standard tool in game theory. His classic works in theoretical biology include *The Evolution of Sex* (1978), *Evolution and the Theory of Games* (1982), and *The Major Transitions in Evolution* (with E. Szatmáry, 1997).

HENRY STAPP is a theoretical physicist at the University of California's Lawrence Berkeley Laboratory, specializing in the conceptual and mathematical foundations of quantum theory, and in particular on the quantum aspects of the relationship between our streams of conscious experience and the physical processes occurring in our brains. He is author of two books on this subject: *Mind, Matter, and Quantum Mechanics* (1993) and *Mindful Universe: Quantum Mechanics and the Participating Observer* (2007).

KEITH WARD is Emeritus Regius Professor of Divinity at Oxford University, Professorial research fellow at Heathrop College, London, and Fellow of the British Academy. Laid out in more than 25 books, his work covers wide areas, from systematic and philosophical theology to comparative theology and science and religion. In 2008 he finished a five-volume series of comparative theology, comparing the great religious traditions on concepts of revelation, creation, human nature, and community. Also in the field of science and religion he has produced a number of influential books, including more recently *Pascal's Fire: Scientific Faith and Religious Understanding* (2006) and *The Big Questions in Science and Religion* (2008).

MICHAEL WELKER held faculty positions at the universities in Tübingen and Münster before he was offered the Chair of Systematic Theology at Heidelberg in 1991.

He has been Director of the University's Internationales Wissenschaftsforum and is currently the Director of the Research Center for International and Interdisciplinary Theology. His method is to work through Biblical traditions as well as through contemporary philosophical, sociological, and scientific theories to address questions of contemporary culture and religion. His influential works in theology include *Creation and Reality* (1999), *What Happens in Holy Communion?* (2000), and *God the Spirit* (2004). Contributions to science and religion include *The End of the World and the Ends of God* (2000) and *Faith in the Living God* (with John Polkinghorne, 2001).

ACKNOWLEDGMENTS

This book grew out of a symposium held in the Consistorial Hall of Copenhagen University on 17–19 August 2006 under the aegis of the John Templeton Foundation and the Copenhagen University Research Priority Area on Religion in the 21st Century. The aim of the conference was to explore fundamental concepts of matter and information in current physics, biology, philosophy and theology with respect to the question of ultimate reality.

We, the editors and co-chairs, arranged the symposium 'God, Matter and Information. What is Ultimate?' in close collaboration with Dr Mary Ann Meyers, the Director of the Humble Approach Initiative under the John Templeton Foundation. The Humble Approach supports cutting-edge interdisciplinary research, insofar as it remains sensitive to disciplinary nuances, while looking for theoretical linkages and connections. Such studies are especially needed in areas of research that are central to the sciences, pertinent for a contemporary metaphysics, and yet are difficult to conceptualize and present in overview.

We are grateful to Mary Ann Meyers for her ongoing enthusiasm and expertise, and to the John Templeton Foundation for sponsoring the symposium so generously. We also want to thank the Editorial Director of Cambridge University Press, Dr Simon Capelin, for his assistance and encouragement in the publication of this book, and the anonymous peer reviewers who supported it. Lindsay Barnes and Laura Clark of the Press have set the editorial standards for this volume and worked in close collaboration with graduate student Trine-Amalie Fog Christiansen at Copenhagen University, who worked as a research assistant on this book and time and again showed her analytical skills. We owe thanks to her, and to Mikkel Christoffersen for assisting in the last phase of the production and for preparing the index.

With two exceptions, all papers grew out of the Copenhagen symposium. We asked Professor Philip Clayton to write a brief philosophical history of the concept of matter, with special emphasis on modernity, and we thank him for doing this so swiftly and well. We also wanted to include the programmatic article of the late evolutionary biologist John Maynard Smith, 'The Concept of Information in Biology' (*Philosophy of Science* 67(2), June 2000); we acknowledge the journal for giving us the permission to reprint this article as Chapter 7 of this volume.

This volume is dedicated to the memory of Arthur Peacocke who, sadly, died on 21 October 2006. Arthur Peacocke was part of the group, but because of his illness he could not attend the conference, so his paper was discussed in his absence. Chapter 12 in this volume is one of the last works from his hand. Peacocke's research in biochemistry and in the intersection of theology and science is highly regarded, and his intellectual testimony can be found in his posthumous *All That Is: A Naturalistic Faith for the 21st Century* (Fortress Press, 2007). But for many of us, Arthur was not just a great scholar, but a mentor, a fellow-inquirer, and a friend who continued to listen, explore, and ask for more. We are indeed indebted to Arthur for his personal combination of rigour and generosity.

Introduction: does information matter?

PAUL DAVIES AND NIELS HENRIK GREGERSEN

~

It is no longer a secret that inherited notions of matter and the material world have not been able to sustain the revolutionary developments of twentieth-century physics and biology. For centuries Isaac Newton's idea of matter as consisting of 'solid, massy, hard, impenetrable, and movable particles' reigned in combination with a strong view of laws of nature that were supposed to prescribe exactly, on the basis of the present physical situation, what was going to happen in the future. This complex of scientific materialism and mechanism was easily amalgamated with common-sense assumptions of solid matter as the bedrock of all reality. In the world view of classical materialism (having its heyday between 1650 and 1900), it was claimed that all physical systems are nothing but collections of inert particles slavishly complying with deterministic laws. Complex systems such as living organisms, societies, and human persons, could, according to this reductionist world view, ultimately be explained in terms of material components and their chemical interactions.

However, the emergence of thermodynamics around 1850 already began to cast doubt on the universal scope

Information and the Nature of Reality: From Physics to Metaphysics, eds. Paul Davies and Niels Henrik Gregersen. Published by Cambridge University Press

of determinism. Without initially questioning the inherited concepts of corpuscular matter and mechanism, it turned out that the physics of fluids and gases in thermodynamically open systems can be tackled, from a practical point of view, only by using statistical methods; the aim of tracking individual molecules had to be abandoned. In what has been aptly called *The Probabilistic Revolution* (Krüger, Daston, and Heidelberger, 1990), determinism became a matter of metaphysical belief rather than a scientifically substantiated position. By the 1870s a great physicist such as James Clerk Maxwell was already questioning the assumption of determinism by pointing to highly unstable systems in which infinitesimal variations in initial conditions lead to large and irreversible effects (later to become a central feature of chaos theory). It was not until the twentieth century, however, that the importance of non-equilibrium dissipative structures in thermodynamics led scientists such as Ilya Prigogine (1996) to formulate a more general attack on the assumptions of reversibility and scientific determinism.

What happened, then, to the notion of matter and the material? In a first phase the term 'matter' gradually lost its use in science to be replaced by more robust and measurable concepts of mass (inertial, gravitational, etc). The story of the transformations of the idea of matter into something highly elusive yet still fundamental is told in detail by Ernan McMullin and Philip Clayton in Chapters 2 and 3 of this volume. Here it suffices to point to three new developments of twentieth-century physics in particular that forced the downfall of the inherited Matter Myth, and led to new explorations of the seminal role of information in physical reality.

2

The first blow came from Einstein's theories of special relativity (1905) and general relativity (1915). By stating the principle of an equivalence of mass and energy, the field character of matter came into focus, and philosophers of science began to discuss to what extent relativity theory implied a 'de-materialization' of the concept of matter. However, as McMullin points out, even though particles and their interactions began to be seen as only partial manifestations of underlying fields of mass-and-energy, relativity theory still gave room for some notion of spatio-temporal entities through the concept of 'rest mass'.

The second blow to classical materialism and mechanism came with quantum theory, which describes a fundamental level of reality, and therefore should be accorded primary status when discussing the current scientific and philosophical nature of matter. In Chapters 4, 5, and 6 Paul Davies, Seth Lloyd, and Henry Pierce Stapp challenge some widely held assumptions about physical reality. Davies asks what happens if we do not assume that the mathematical relations of the so-called laws of nature are the most basic level of description, but rather if *information* is regarded as the foundation on which physical reality is constructed. Davies suggests that instead of taking mathematics to be primary, followed by physics and then information, the picture should be inverted in our explanatory scheme, so that we find the conceptual hierarchy: information → laws of physics → matter. Lloyd's view of the computational nature of the universe develops this understanding by treating quantum events as 'quantum bits', or qubits, whereby the universe 'registers itself'. Lloyd approaches this subject from the viewpoint of quantum information science, which sets as a major goal the construction of a quantum computer – a device that can process information at the quantum level, thereby achieving a spectacular increase in computational power. The secret of a quantum computer lies with the exploitation of genuine quantum phenomena that have no analogues in classical physics, such as superposition, interference, and entanglement. Quantum computation is an intensely practical programme of research, but Lloyd uses the concept of quantum information science as the basis for an entire world view, declaring that the universe as a whole is a gigantic quantum computer. In other words, nature processes quantum information whenever a physical system evolves.

Lloyd's proposal forms a natural extension of a long tradition of using the pinnacle of technology as a metaphor for the universe. In ancient Greece, surveying equipment and musical instruments were the technical wonders of the age, and the Greeks regarded the cosmos as a manifestation of geometric relationships and musical harmony. In the seventeenth century, clockwork was the most impressive technology, and Newton described a deterministic clockwork universe, with time as an infinitely precise parameter that gauged all cosmic change. In the nineteenth century the steam engine replaced clockwork as the technological icon of the age and, sure enough, Clausius, von Helmholtz, Boltzmann, and Maxwell described the universe as a gigantic entropy-generating heat engine, sliding inexorably to a cosmic heat death. Today, the quantum computer serves the corresponding role. Each metaphor has brought its own valuable insights; those deriving from the quantum computation model of the universe are only just being explored.

In the absence of a functional quantum computer, the most powerful information-processing system known is the human brain (that may change soon, as even classical computers are set to overtake the brain in terms of raw bit flips). The relationship between mind and brain is the oldest problem of philosophy, and is mirrored in the context of this volume by the information–matter dichotomy. Crucially, the brain does more than flip bits. Mental information includes the key quality of semantics; that is, human beings derive understanding of their world from sense data, and can communicate meaning to each other. The question here is what can, and what cannot, be explained merely by digital information, which is formulated in terms of bits without regard to meaning. When the foundation for information theory was laid down by Shannon, he purposely left out of the account any reference to what the information means, and dwelt solely on the transmission aspects. His theory cannot, on its own, explain the semantics and communication of higher-order entities. At most, one could say, as Deacon suggests in Chapter 8, that Shannon focused on the syntactic features of an information potential.

The foregoing properties of the mental realm are closely related to the issue of consciousness. How the brain generates conscious awareness remains a stubborn mystery, but there is a well-established school of thought that maintains it has something to do with quantum mechanics. Certainly the role of the observer in quantum mechanics is quite unlike that in classical mechanics. Moreover, if quantum mechanics really does provide the most fundamental description of nature, then at some level it must incorporate an account of consciousness and other key mental properties (for example, the emergence of semantics, the impression of free will). For many years, Henry Stapp has championed the case for understanding the mind and its observer status in a quantum context, and in Chapter 6 he sets out a well-argued case both for taking consciousness seriously (that is, not defining it away as an epiphenomenon) and for accommodating it within a quantum description of nature.

The third challenge to the inherited assumptions of matter and the material comes from evolutionary biology and the new information sciences, which have made revolutionary discoveries since the 1940s and 1950s. Placed at the interface of the physical and cultural sciences, biology plays a pivotal role in our understanding of the role of information in nature. In Chapter 7 John Maynard Smith argues that the biological sciences must be seen as informational in nature, since the sequence structure of DNA is causally related, in a systematic way, to the production of proteins. In the nineteenth century, living organisms were viewed as some sort of magic matter infused with a vital force. Today, the cell is treated as a supercomputer – an information-processing and -replicating system of extraordinary fidelity. The informational aspects of modern molecular biology are conspicuous in the way that gene sequencing and gene pathways now form the foundation for understanding not only evolutionary biology, but also cell biology and medicine. In Chapters 8 and 9 Terrence Deacon and Bernd-Olaf Küppers offer two distinct naturalistic views about how the crucial *semantic* levels of information might emerge via

6

thermodynamical (Boltzmann) and evolutionary (Darwinian) processes. Both accounts argue that biological information is not only instructional but also has to do with 'valued' or 'significant' information, which puts the receiver in the centre of interest. Significant information, however, is always a subset of a wider set of informational states, which may be described as the underlying 'information potential'. With this background, Deacon presents a naturalistic theory of the emergence of contextual information; that is, the capacity for reference and meaning, which he describes in terms of the notion of 'absent realities'. This he accomplishes by combining the Shannon–Boltzmann view that information is always relative to a statistical information potential, with the Darwinian emphasis on what actually works for an organism in its pragmatic setting. In Chapter 10 Jesper Hoffmeyer then presents a biosemiotic proposal, which questions the overarching role of genetics, and rather opts for the importance of a cell-centred view. Finally, in Chapter 11, Holmes Rolston offers a natural history of the emergence of an informed concern for others. Evolution is a notoriously 'selfish' process, but eventually it generates systems that display altruism and exhibit concern for other beings. With the increase of sense perception and the top-down capacities of mammalian brains, an ethical dimension of nature arrives on the evolutionary scene. A cell-centred view is not necessarily a self-centred view.

It would be wrong to claim that the science-based chapters collectively amount to an accepted and coherent new view on the fundamental role of information in the material world. Many scientists continue to regard matter and energy as the primary currency of nature, and information to be a secondary, or derived concept. And it is true that we lack the informational equivalent of Newton's laws of mechanics. Indeed, we do not even possess a simple and unequivocal physical measure for information, as we have for mass and energy in terms of the units of *gram* and *joule*. Critics may therefore suspect that 'information' amounts to little more than a fashionable metaphor that we use as a shorthand for various purposes, as when we speak about information technologies, or about anything that is 'structured', or some way or another 'makes sense' to us.

The incomplete nature of information theory is exemplified by the several distinct meanings of the term 'information' used by the contributors in this volume. Quantum events as informational qubits (Lloyd), for example, have a very different character from Shannon-type digital information, or as mere patterns (Aristotelian information), and none of the foregoing can much illuminate the emergent concept of *meaningful* information (semantic information). In spite of the tentative nature of the subject, however, two reasons can be offered for giving information a central role in a scientifically informed ontology. The main point is that information makes a causal difference to our world – something that is immediately obvious when we think of human agency. But even at the quantum level, information matters. A wave function is an encapsulation of *all that is known* about a quantum system. When an observation is made, and that encapsulated knowledge changes, so does the wave function, and hence the subsequent quantum evolution of the system. Moreover, informational structures also play an undeniable causal role in material constellations, as we see in, for example, the physical phenomenon of resonance, or in biological systems such as DNA sequences. What is a gene, after all, but *a set of coded instructions* for a molecular system to carry out a task? No evolutionary theory can have explanatory function without attending to the instructional role of DNA sequences, and other topological structures. But neither can a bridge or skyscraper be constructed successfully without paying due attention to the phenomenon of resonance, and so it seems that just as *informational events* are quintessential at the lowest level of quantum reality, so are *informational structures* quintessential as driving forces for the historical unfolding of physical reality.

The philosophical perspectives of a material world based on an irreducible triad of mass, energy, and information are discussed in the contributions in the section on philosophy and theology. In Chapter 12 the late biologist and theologian Arthur Peacocke (to whom this book is dedicated) presents his integrative view about how an emergentist monism, informed by the sciences of complexity, must be sensitive to the uniformity of the material world as well as to the distinctive levels that come up at later stages of evolution. Peacocke's theological synthesis thus combines naturalism and emergentism with a panentheistic concept of God; that is, God permeates the world of nature from within, although God is more than the world of nature in its entirety. Peacocke's religious vision is thus developed within the horizon of what he calls EPN (emergentist/monist–panentheistic–naturalist). In Chapters 13 and 14 the philosophical theologians Keith Ward and John F. Haught explore novel ways for

9

understanding God as the source of information for a self-developing world. Ward argues for what he calls a supreme informational principle of the universe, without which the combination of the lawfulness of the world and its inherent value would be inexplicable. Such informational code for construction of an actual universe logically precedes material configurations by containing the set of all mathematically possible states, plus a selective principle of evaluation that gives preference to the actual world that we inhabit. Ward suggests that this primary ontological reality may be identified with God, especially if the given laws of nature can be seen as providing space for qualities such as goodness and intrinsic value. Haught argues that information must walk the razor's edge between redundancy (too much order) and noise (too much contingency). It is this felicitous blend of order and novelty that transforms the universe from a mere physical system into a narrative of information processing. While reminding us that all 'God language' must be regarded as analogical, he argues that the concept of God as an informational principle at work in the entire cosmic process is far richer than the idea of a designer God at the edge of the universe. While emphasizing the logical space of all nature (Ward) and the evolutionary unfolding thereof (Haught), both draw on contemporary scientific accounts of nature that accord with, or even suggest, a divine reality with world-transforming capacities. A science-based naturalism may thus still allow a distinction between the world of nature (with a small 'w') and the World in extenso (with a capital 'W'). Finally, in Chapters 15 and 16, Niels Henrik Gregersen and Michael Welker argue that the new scientific perspectives of matter and information summarized in this volume give fresh impetus to a reinterpretation of important strands of the Biblical traditions. Gregersen shows how the New Testament concept of a 'divine Logos becoming flesh' (John 1:14) has structural similarities to the ancient Stoic notion of Logos as a fundamental organizing principle of the universe, and should not prematurely be interpreted in a Platonic vein. The Johannine vision of divine Logos being coextensive with the world of matter may be sustained and further elucidated in the context of present-day concepts of matter and information, where the co-presence of order and difference is also emphasized. A typology of four types of information is presented, reaching from quantum information to meaning information. In the final essay, Welker suggests that interdisciplinary discussions (between science, philosophy, and theology) should be able to move between more general metaphysical proposals and the more specific semantic universes, which often are more attentive to the particulars. One example is Paul's distinction between the perishable 'flesh' and the possibility of specific 'bodies' being filled with divine energy. Such distinctions may also be able to catch the social dimensions of material coexistence, which are left out of account in more generalized forms of metaphysics. According to Paul, the divine Spirit may saturate the spiritual bodies of human beings and bring them into communication, when transformed in God's new creation.

Our hope is that the selection of essays presented in this volume will open a new chapter in the dialogue between the sciences, philosophy, and theology.

References

Krüger, L., Daston, J., and Heidelberger, M., eds (1990). *The Probabilistic Revolution*, Cambridge, MA: MIT Press.

Prigogine, I. (1996). *The End of Certainty. Time, Chaos, and the New Laws of Nature*, New York: The Free Press.

PART I
HISTORY

~

From matter to materialism... and (almost) back

ERNAN McMULLIN

~

The matter concept has had an extraordinarily complex history, dating back to the earliest days of the sort of reflective thought that came to be called 'philosophy'. History here, as elsewhere, offers a valuable means of understanding the present, so it is with history that I will be concerned – history necessarily compressed into simplified outline.

This story, like that of Caesar's Gaul, falls readily into three parts. First is the gradual emergence in early Greek thought of a factor indispensable to the discussion of the changing world and the progressive elaboration of that factor (or, more exactly, cluster of factors) as philosophic reflection deepened and divided. Second is the radical shift that occurred in the seventeenth century as the concept of matter took on new meanings, gave its name to the emerging philosophy of materialism and yielded place to a derivative concept, mass, in the fast-developing new science of mechanics. Third is the further transformation of the concept in the twentieth century in the light of the dramatic changes brought about by the three radically new

Information and the Nature of Reality: From Physics to Metaphysics, eds. Paul Davies and Niels Henrik Gregersen. Published by Cambridge University Press
© P. Davies and N. Gregersen 2010, 2014.

theories in physics: relativity, quantum mechanics, and expanding-universe cosmology, with which that century will always be associated. Matter began to be dematerialized, as it were, as matter and energy were brought into some sort of equivalence, and the imagination-friendly particles of the earlier mechanics yielded way to the ghostly realities of quantum theory that are neither here nor there.

2.1 Phase one: origins

Aristotle was the first to adapt a term from ordinary usage in order to anchor the technical analysis of change that is at the heart of his physics (Johnson, 1973; McMullin, 1965). *Hylē* (rendered as *materia* in Latin) originally designated the timber used, for example, in the construction of a house, the materials of a making. But long before Aristotle, the very first Greek philosophers sought to specify the 'stuff' from which they believed all things to have originally come, a stuff that would have familiar sense-properties that would help to make cosmic origins intelligible. Some went on to intimate that this is also the stuff of which all things are now composed. They disagreed as to what that stuff might be – water, fire, atoms, seeds, ... – and they lacked any generic term for the stuff itself. But the notion that there should be something or other that underlay not only origins but perhaps all change was a key feature of the thinking of these first Ionian 'philosophers'.

Nor did Plato later have a specific term to play the role, for him quite central, that would later be assigned to matter (Eslick, 1965). The search that forever occupied him was for intelligibility, and this he found in the first instance in mathematics, where Reason can discern Forms that are beyond the reach of change, the enemy of intelligibility. The world of sense is a changing world, and by that very fact it is for Plato only imperfectly intelligible. Forms are instantiated there; but the Form of a triangle, say, is only imperfectly realized in the multiplicity of objects in the sense world that can pass as triangular. What makes this sort of multiplicity possible is the presence of a Receptacle, existent in its own right, somewhat akin to what we would call 'space'.

The Receptacle is what makes the sense world possible. It instantiates an indefinitely great number of individuals that participate in the Forms, and in instantiating it also individuates. However, the Receptacle is unstable; it has a 'shaking motion' that prevents the Forms from being fully realized within it. The world of sense is thus at best a world of image, not of true reality, which resides only in the ideal world of Form. The matter receptacle is thus the source of the manifold defects of the world of sense, of the many ways in which it falls away from intelligibility and thus from the Good. The division between matter and mind is absolute.

In one crucial respect, Aristotle stood the system of his mentor, Plato, on its head. For him, change is not necessarily a falling away from intelligibility: it is the very means through which the intelligibility of the world around us is to be discerned. The paradigm for him was the living world, not the world of mathematical Form. He looked to behaviour – that is, change – to understand what a nature is, to what kind it belongs. Mathematical forms may be imperfectly realized by wooden triangles or copper squares, but living forms, in contrast, are ordinarily fully realized in the individual member of the species.

Aristotle began from the logic of change-sentences (McMullin, 1965). In contrast to Parmenides, who attempted to generate paradox by construing change as a movement from non-Being to Being, Aristotle saw such sentences as triadic: a subject (a leaf, say), a predicate (brown), and a lack, to begin with, of that predicate (non-brown). This allowed him to point out that the original lack cannot be regarded as non-Being; it conveys something real, namely the capacity or potentiality to become brown, a distinctive fact about the leaf. The 'matter' of the change is that which remains throughout – but it also, crucially, is the bearer of potentiality. Green leaves, though not brown, have the distinctive potentiality to turn brown at some point. This is something real about leaves, a potentiality that is crucial to our understanding of what leaves are. This important moral is one that may be drawn from every sentence that denotes a change.

The world for Aristotle was made up of substances, unities that exist in their own right as living things do. What if one of these ceases to be? What occurs is a change, not a replacement, so something must remain the same. The matter of such a change cannot have properties of any kind, he argued, for if it does, it must itself be a substance – which would mean that the change itself in that case would not be a truly substantial one. (This is his main objection to the various sorts of underlying 'stuff' postulated by his Ionian predecessors.) This led him to introduce first, or 'prime', matter, something that lacks all properties of its own and functions only to ensure the continuity, and hence the reality, of substantial change.

Aristotle himself had little to say about prime matter, but his successors invested much effort in attempting to clarify this controversial notion.

What stands out in this story is that matter, as the one who introduced the term understood it, is not distinguished by any particular property. In the case of ordinary (non-substantial) change, the 'matter' ('second' matter, as it came to be called) is just the subject of the change, whatever its properties happen to be (a leaf, for instance). Where the change is substantial, the matter is necessarily property-less, not a constituent in the ordinary sense but something that came to be described as a metaphysical principle (McMullin, 1965, pp. 173–212). But whether it possesses formal properties or not, matter was seen by Aristotle first and foremost as the bearer of potentiality.

There were other features of the matter concept that came in for fuller treatment in the later medieval tradition. One was individuation (Bobik, 1965). Aristotle's substances had form and matter as co-principles or aspects, each requiring the other in order to constitute an existent thing. Form obviously could not individuate; it could be instantiated indefinitely many times. So individuation would have to come from the side of matter, and prime matter at that. But how? Prime matter was supposed to be indeterminate. Individuation clearly had something to do with location in space and time. After much discussion, it was decided that these properties belonging to the Aristotelian category of quantity would have to be part, at least, of the thing that designated something as an individual: '*materia signata quantitate*', as the phrase went. It was not at all clear that this new role attributed to matter was compatible with the earlier account of substantial change, although individuation clearly was involved in the ensuring of continuity of a body through such change.

What should be noted, however, is the introduction at this point of the notion of a 'quantity of matter' (Weisheipl, 1965). That would bear fruit in a different context – in the study, for example, of the phenomena of rarefaction and condensation, taken to be straightforward examples of substantial change of one element (water, say) into another ('air', or vapour). There are clearly quantitative constraints on such a change; the quantity of some 'stuff' otherwise indeterminate must be conserved. Richard Swineshead suggested, on intuitive grounds, that the quantity of this stuff, the quantity of matter, should be proportionate to the volume as well as to the density of the body concerned: the definition that Newton would later adopt as his own.

Parallel to this but in the very different context of motion, Jean Buridan postulated an 'impetus' in the case of moving bodies that is conserved in unimpeded motion and is a measure of resistance to change of motion. He too made it proportionate to the body's quantity of matter. The quantification of matter in this context was obviously a major step towards the mechanics of a later age. The issue of conservation had pointed Aristotle to the concept of matter in the first place, but now it was leading in a very different direction, one he had surely not anticipated.

One other development owed more to Plato than to Aristotle. From the beginning, Christian theologians favoured the strong dualism of body and soul that characterized the neo-Platonic tradition. This dualism was also described as being between 'matter' and 'spirit', thus leading to another rather different sense of the term matter – namely as a generic term that describes any item in the physical world. To be 'material' in this sense is simply to belong to the physical world, the contrast here also often described as being between the corruptible and the incorruptible. The operation of the human intellect was taken to be of a 'non-material' sort, incapable of being reduced to 'material' categories; indeed (as Aquinas and others held), ultimately independent of the operation of the brain. Defining mind in terms of the immaterial or the non-material would leave behind the issue of how the boundaries of 'material' action were to be drawn. No one at that earlier time could have guessed how difficult that would later become.

2.2 Phase two: transformation

The seventeenth century marked a transformation in the concept of matter; one in which the burgeoning science of mechanics played the principal role. With the shift of focus from the world of living things to the more generic topic of bodies in motion came the rejection of the Aristotelian category of substance that had depended so much on the organism as paradigm. And with the disappearance of sub-stantial form came the removal of the barrier to regarding change as involving 'stuff' with specific properties. In this reversal, Descartes played an important part (Blackwell, 1978). Convinced that the world should be fully intelli-gible to the human mind, and convinced further that the intelligibility of geometry furnished the model he needed, he equated the stuff of which the world was made – its 'matter' – with extension. Reducing the matter of bod-ies to their extension, a combination of their volume and their shape, would make the world fully amenable to the methods of geometry. The science of motion could then be rendered entirely mathematical, with the help of two intuitive principles: conservation of motion; and restriction of action between bodies to contact only.

However, there were some obvious barriers in the way of such a reductive picture of matter. First, the property of impenetrability would have to be smuggled in: extensions as such cannot collide! Second, and more serious, the lack of anything corresponding to density would make the construction of a plausible mechanics very difficult, if not impossible. Bodies of different densities obviously have different mechanical properties. Third, as matter and extension are the same, there is no empty space. How then are bodies to move? Descartes displayed extraordinary ingenuity in an effort to get around these and other difficulties, but it eventually became clear that a matter with only the single property of extension could not furnish the basis for an adequate mechanics.

The incentive for such a reduction remained, however, although now more realistically moderated to admit a handful of properties besides shape and volume: impenetrability, mobility, inertia, and perhaps density. There was a way of getting round admitting this last property: if one adopted the corpuscular model of matter, density could be explained in terms of degree of packing of corpuscles (assumed to be of uniform density) in an otherwise empty space. These properties were often defined as the 'primary' qualities of matter, in contrast to the 'secondary' ones. 'Primary' here could mean objective rather than subjective qualities, understanding the latter as dependent in one way or another on a perceiver. Or it could refer to the qualities in terms of which all others could be explained (McMullin, 1978a, pp. 32–38).

Either way, this commonly drawn distinction supported the 'mechanical philosophy' that came to dominate the work of seventeenth-century natural philosophers. According to these philosophers, matter was characterized by a small set of properties drawn from everyday experience. It was composed of tiny corpuscles, themselves invisible. The argument for a claim so sharply at odds with the growing empiricism of the day depended primarily at this point on a widely (although not universally) shared repugnance with regard to the consequences of allowing matter to be divisible into an infinite number of parts (Holden, 2006).

Corpuscular matter (so it was hoped) could in principle explain all the changes taking place in the visible world: for example, the chemical transformations described by Boyle in terms of underlying corpuscular structures and motions. As some saw (but others did not), validating explanations of this sort would require a new form of inference, one relying in the first instance on imagination for the invention of hypotheses about these underlying structures and requiring criteria for the assessment of a hypothesis more sensitive than a mere saving of the phenomena (McMullin, 1994). At this early stage such a programme was still largely promissory, of course. The unbounded faith in the corpuscular hypothesis was not matched by concrete results: one can point to scarcely a single successful explanation in corpuscular terms from that century.

The reductivist agenda of the mechanical philosophy fitted well with the growth, particularly in France, of what would come to be called 'materialism', that label recalling the neo-Platonic contrast between matter and spirit. In the writings of La Mettrie, d'Holbach, and their followers, materialism would take on a variety of forms, but it would always involve a denial of the existence of spirit, understood as something that lay outside the scope of the mechanical philosophy, and it would emphasize the reductivist claims of that philosophy. The 'matter' of the materialism of that day was made up in the first instance of the bodies of our ordinary experience, themselves believed to be composed of smaller bodies, like the larger ones in all respects save size. And the new mechanics would suffice (Newton postulated) for the small as it already had for the large.

Mentioning Newton recalls one further shift in the concept of matter, a shift for which he was responsible (McMullin, 1978b). He needed in his new mechanics a measure of a body's resistance to change in motion, as well as a body's gravitational effect on other bodies and its own capacity to be acted upon gravitationally. A fundamental postulate of his new mechanics was that a single quantity would be the measure of all three, plausibly, the amount of 'stuff' or the 'quantity of matter' (Buridan's term) of the body. For this, Newton introduced a convenient abbreviation, 'mass'. At that point, the older term 'matter' ceased to have any explicit function in the new mechanics, although it could still be said to remain implicitly present as the bearer of mass, whether inertial or gravitational. Matter could now be defined quite simply as that which possesses mass.

But what was one to make of the gravitational attraction at the heart of the new system? It clearly lay outside the bounds of the mechanical philosophy as it had been understood up to that time. It appeared to pose an unwelcome decision: either to treat it as action at a distance (generally regarded as unacceptable) or as mediated by something across intervening space. Newton himself had showed that this latter could not be a medium possessing inertial mass. But what other alternative was there? Over the years following the publication of the *Principia*, he tried out a variety of ideas, among them an 'elastic spirit' and an 'immaterial' medium (McMullin, 1978b, pp. 75–109). The reductivist concept of matter of the mechanical philosophers was clearly coming under strain at this point.[1]

Furthermore, matter itself seemed to occupy an ever-smaller part of Newton's universe. How were the particles of which light appeared to be composed able to traverse transparent media of considerable thickness? If they struck material corpuscles, they would surely be halted or at least diminished. Transparency could be explained only by postulating that the material corpuscles or atoms occupied merely a tiny part of the transparent body. And if this were so, might it not be the case in non-transparent bodies too? In the opinion of Samuel Clarke, Newton's disciple, the new mechanics entailed the view that matter is 'the most inconsiderable part' of the universe; the immaterial forces that in effect filled space were what really counted. Joseph Priestley would later write that according to the 'Newtonian philosophy' all the solid matter in the solar

[1] Leibniz was an important contributor to this discussion; see Chapter 3 by Philip Clayton in this volume.

system 'might be contained within a nutshell' (Thackray, 1968).

Roger Boscovich carried this thought to its logical conclusion, dismissing solid extended matter entirely and replacing it with point-centres of force. Besides the long-range gravitational force of attraction, he postulated a short-range force of repulsion that would constitute something like an extended atom around each point-centre. This 'atom' would not have a well-defined surface but the force of repulsion would make penetration more and more difficult, and in the end physically impossible, as the centre was approached.

Before this, potentiality had always been associated with an actuality of some sort. But Boscovich's atoms were not actual; they were themselves potential only, specifying what would happen if... If what? Could a potentiality trigger another potentiality? Clearly, a new sort of status was being attached to potentiality here, a kind of shadowy materiality. Others, like Kant in his early writings, would take up the challenge offered by this apparent paradox, noting that it at least avoided the more familiar paradoxes associated with infinite divisibility (Holden, 2006). However, the stubborn actuality of the world of ordinary experience made it difficult to countenance so radical a move.

And this sort of challenge would steadily strengthen as the eighteenth and nineteenth centuries progressed. Static electricity could apparently be stored. Ought it, therefore, to qualify as 'material'? The successes of the wave theory of light suggested an intervening medium of some sort... but what sort? Newton had shown that such a medium could not be inertial. But in that case, how could it be material? What was now to count as 'matter'?

It was from electromagnetic, not gravitational, theory that a resolution would ultimately emerge (McMullin, 2002). The successful portrayal of electromagnetism in field terms convinced Maxwell that the field had to be regarded as something more than a convenient calculational device. It had to designate a reality of a new sort, defined by the energy it carried, although understood as pure potentiality. In the earlier days of Newtonian mechanics, energy and momentum only gradually came to be recognized as significant quantities in their own right, mass retaining its dominant role. For a time, it was not clear which of the two, energy or momentum, would prove to have the all-important property of being conserved through change. Finally, that role was bestowed on energy. And with the advent of the field concept, energy did indeed appear to be emerging as a reality in its own right. But how was it to be related to matter? An entirely unexpected answer to that question was soon to make its appearance.

2.3 Phase three: dematerialization?

The physics of the twentieth century led the human imagination into domains stranger and stranger, domains as remote as could be from the comfortably familiar matter of yesteryear. In that respect, the shift might be described as a progressive dematerialization, each stage corresponding to one of the three great advances that physics made in the course of the new century.[2]

---

[2] I am particularly grateful to Gerald Jones and William Stoeger S.J. for their unstinting help in illuminating some of the more technical issues discussed in this section, and to Paul Davies for a valuable correction.

2.3.1 Relativity

The special theory of relativity ('special' because it deals only with the special case of uniform motion) came as the first warning shock of a radically new era in physics. Its farthest-reaching consequence was assuredly the mass–energy equivalence described in Einstein's famous equation: $E = mc^2$. It implies that rest mass can be transformed into radiation, and vice versa. In a process in which mass is lost (as with certain kinds of fission or fusion at the atomic level), the mass deficit is made up by the release of energy in the form of (massless) radiation as well as the kinetic energy of the final-state particles (if any) involved. Conversely, energy can be transformed into mass in the process of pair creation when the energy available is high enough to transform into the rest masses of the two particles. A gamma-ray (high-energy) photon can create an electron–positron pair if it strikes a nucleus that absorbs the photon's momentum; two such photons in collision in the presence of an electrical charge likewise can create a pair of particles if their joint energies are high enough to supply the two rest masses needed.

This is a startling demotion of matter as the sole carrier of the 'reality' label. Something without mass is, by the Newtonian definition at least, something without any quantity of matter. Massless radiation would not, then, qualify as matter. The Einstein equivalence equation has, in effect, begun the 'dematerialization' of physical reality. The only way in which the world can still be described as the 'material' world, or the term materialism can preserve its original significance, is to redefine 'matter'. But how? Materialism, if one wants to retain the term, seems to have unexpectedly become a much more open doctrine.

In his general theory of relativity, Einstein went on to consider the more complicated case of accelerated motion, and hence of force. Here, the mass–energy equivalence has a further consequence. Whereas in the past mass was the sole measure of gravitational agency, both active and passive, that role is now transferred to mass–energy. In the Einstein field equations, the stress-energy tensor on the right-hand side describes the sources of the gravitational field, and these now include energy, both kinetic and potential, as well as rest mass. These can be summed together with the aid of the equivalence. So a photon, for example, despite having zero rest mass, will exert gravitational force in virtue of its kinetic energy. And a tightly wound watch will contribute slightly more to gravity in virtue of its potential energy than one that has run down.

When Newton defined mass as 'the quantity of matter' for the purposes of his mechanics, he had in mind its inertial and gravitational roles. These roles had by now been enlarged by the two theories of relativity, and mass–energy had taken the place of mass alone in relativistic mechanics.[3] However, rest mass and energy, although physically convertible into one another, still retain their separate identities: rest mass is not a form of energy, and

---

[3] It should be noted that in all of this, 'mass' is to be understood as 'rest mass.' An alternative, less popular, usage would redefine 'mass' in relativistic terms to include kinetic and potential energy as well as rest mass. According to this definition, photons would have 'mass' ('relativistic' mass) and the gravitating entity would once again be 'mass' only, but now in an altered sense: one that obscures the still fundamental difference between rest mass and energy.

energy does not have rest mass. When the gravitating entity here is described as mass–energy, the term thus refers to the composite: rest mass plus energy. ('Energy–momentum', the alternative label, is less explicit about the continuing role of matter, our topic here.)

Instead of bodies being said to affect one another by the exercise of gravitational force, mass–energy is now said to 'warp' the space–time in its neighbourhood, as described by the non-Euclidean metric and its derivatives on the left side of the field equation. Replacing Newton's force, which acts mysteriously at a distance, with Einstein's mass–energy, which 'causes' the local 'warping' of a no longer Euclidean space–time, might be thought a somewhat dubious gain in terms of understanding! One bonus, at least, is that the equality of the inertial and the gravitational measures of the new mass–energy entity now follows from the theory itself and thus does not require the further empirical support it needed in the Newtonian system.

If the term 'matter' is to be retained at all in the midst of these fundamental changes in our understanding of that of which our world is made, two alternatives appear to be open. It could be broadened to include mass–energy, which now has the role that mass once had in mechanics, or it could still be restricted to rest mass, leaving the world with two constituents – matter and energy – rather than one.

2.3.2 Quantum theory

A second theory that did even more violence to everyday convictions was the quantum theory of the 1920s. Here, once again, the concept of matter was at the forefront of the change brought about by the theory. Ever since the time of Newton and Huygens, the physics of light had wavered between two poles: one representing light as corpuscular, the other as periodic, or wave-like. By the beginning of the twentieth century, it became clear that light would have to be regarded somehow as both. This was all very well, perhaps, for light, which had always had an ethereal quality anyway. But matter? Surely not matter? The wave behaviour of matter at the subatomic level first postulated on theoretical grounds by de Broglie and later confirmed observationally by Davisson and Germer perhaps came as an even more severe shock to common-sense intuitions than the one they had already suffered from the theories of relativity.

Let us return for a moment to the 'matter' to which the classical materialisms of earlier centuries made reference. It was defined in terms of a few primary qualities, each of which had the virtue of being intelligible in terms of everyday experience. Extension, shape, density, motion, impenetrability . . . everyone knew what they amounted to. There was no need to refer to an abstract theory in order to situate them. In his crucial third rule of reasoning, Newton had postulated as a regulative principle of physical science generally that, in effect, the very small and the very distant should have the same accessible properties as those familiar to us in everyday experience (McMullin, 1992).

Quantum mechanics set all of that aside. According to this theory, at the level of the very small, the position and the motion of a body cannot both be sharply determined at the same time. (The quantum phenomena become appreciable, in effect, only at the level of the very small.)

According to the majority interpretation of the new theory – the so-called 'Copenhagen' interpretation – electrons do not even possess sharp positions or momenta prior to measurement. (A minority view – the 'Bohm' interpretation – still not definitively set aside, would deny this.) They would not, then, qualify as matter, were the classical seventeenth-century type of definition to be maintained. Roughly speaking, electrons seem to travel as waves but to interact as particles.

Even worse, the quantum uncertainty principle applies not only to the position–momentum pair but also to energy–time. It follows that the energy state of the electromagnetic vacuum fluctuates around the zero value, the maximum amount of such a fluctuation in energy being correlated to the maximum time it may persist. (Strictly speaking, what the uncertainty principle entails, as Heisenberg envisaged it, is only that such a fluctuation, were it to occur, could not be observed, not that such fluctuations actually occur. But no matter...) The fluctuation is then represented as a 'particle' (a photon) of a certain energy. Since, according to the uncertainty principle, it cannot be observed, it has come to be called a 'virtual' particle. The energy added by its sudden appearance would violate conservation of energy were it to be real, but being only virtual, it is granted a (convenient!) exemption from the conservation requirement.

Although virtual particles cannot, in principle, be observed, the net effect of the quantum energy fluctuations in the vacuum *has* been observed, it is claimed; its magnitude was in fact predicted in advance (the Casimir effect). The effect is tiny, requiring a very demanding level of accuracy in a complex experimental arrangement. This has encouraged the conclusion that the virtual particles (in this case, virtual photons, supposedly infinite in number) are nonetheless 'real' in the sense that 'their indirect effects are measurable' (Krauss, 1989, p. 35). 'No doubt remains that virtual particles are really there' (Barrow and Silk, 1993, pp. 65–66). Whether this constitutes an adequate test of 'reality' is disputed. Others insist that in-principle observability is the ineliminable criterion of what counts as a 'real' particle. At this point, the distinction between 'real' and 'virtual' evidently begins to thin out.

'Virtual' particles make an appearance also in a different quantum context. In order to calculate the interactions between (real) particles in quantum field theory, Feynman introduced a simplified diagrammatic method of estimating the relevant probabilities. The forces involved are represented as the exchange of virtual particles between the real particles, carrying the momentum from one to the other that affects their individual motions. This is then said to resolve the difficulty posed by the apparent action at a distance implicit in Newton's notion of gravitational force. Exchange of momentum via the action of virtual particles that travel from one body to the other at last provides an explanation of gravitational action, so it is claimed, and allows one to dispense with the troublesome notion of force entirely. (Whether this really is an improvement in terms of explanation is not entirely obvious.)

Among physicists themselves, there is an evident division as to the status that should be accorded to this sort of 'virtual' entity (Davies, 2006, p. 74). For example: 'Virtual particles are a language invented by physicists in order to talk about processes in terms of Feynman diagrams ... Particle physicists talk about [decay] processes as if the [virtual] particles exchanged . . . are actually there, but they are really only part of a quantum probability calculation . . . they cannot be observed' (SLAC, 2006). Yet the metaphor of particle exchange has clearly taken hold in ways that suggest that, for the average physicist, it is something more than a convenient calculational device.

2.3.3 Cosmology

Moving finally to the fast-growing science of cosmology, it too has cast up matter-related surprises of late. After the big-bang theory of cosmic origins received crucial confirmation in the 1960s, the question of the fate of the universe – continuing expansion or eventual contraction – soon became a focus of attention. It turned out that the balance between the two options was an extremely delicate one: the cosmic mass–energy density today had to be in the neighbourhood of a specific ('critical') intermediate value for runaway expansion or rapid collapse to have been avoided, that is, for the universe to be the long-lived one that it evidently is. (Another way of putting it is that the ratio of the actual value to the critical value, symbolized by the Greek letter omega (ω), had to be not too far from unity, ensuring an approximately 'flat' space–time.) The problem was that for this to be the case, the density would have had to be poised close to the critical value, to a fantastic degree of precision, before the cosmic expansion began, because the effect of the expansion would have been to magnify very rapidly any initial deviation from the critical value. This set off the first round in a celebrated debate about what came to be called 'fine tuning' (McMullin, 2008).

In 1981, Alan Guth proposed an ingenious modification of the big-bang theory that involved a gigantic cosmic inflation in the first fraction of a second of the expansion. Over time, the idea proved to have far-reaching consequences, the first of these being a solution of the critical density problem: the inflation would inevitably push the cosmic curvature towards flatness and hence the mass–energy density towards the critical value.[4] To the extent that the theory of inflation is confirmed, the search for an answer to the cosmic curvature question is answered: it is very close to flat.[5] But that now leaves a rather different question: assuming that the ratio of the actual mass–energy density to the critical value is, in fact, very close to unity, where is that amount of mass–energy to come from?

The trouble is that the 'baryonic matter' of which stars and planets (and ourselves) are made does not come even close to bringing the cosmic mass–energy density up to the critical value; its contribution appears to be no more than perhaps 4% of the desired amount. (What is now routinely called 'matter' here includes radiation and the kinetic and potential energies of all known particles, conforming to the revised definition of 'matter' proposed above.) Enter a new sort of matter: called 'dark' originally because it does not interact with the radiation that is the normal indication of matter's presence, but now rather more because its

---

[4] In that regard, the repulsive 'gravity' of inflation has the opposite effect to that of the attractive gravity involved in gradually slowing the motions consequent to the original 'big bang'.

[5] Recent measurements of small anisotropies in the cosmic microwave background by the Wilkinson Microwave Anisotropy Probe (WMAP) satellite offer strong confirmation for this important claim.

nature is not at all understood despite the most intensive investigation.

As early as the 1930s it was noted that the motions of galaxies within galactic clusters was more rapid than the amounts of conventional gravitating matter could explain. Many suggestions were forthcoming, among them the possibility that Newton's equations might have to be modified in order to apply over such enormous distances. But the theory that eventually gained wide acceptance was that the galaxies were surrounded by vast clouds of something or other that betrays its presence by a single property only: it exerts gravitational attraction. It is believed to have rest mass. More recently, its effects have been identified in several other contexts: gravitational lensing, and the temperature of the X-ray-emitting gases in galactic clusters, for example.

Nevertheless, despite extended efforts, the composition of dark matter still defies identification. For one thing, the relative cosmic abundances of the lightest elements, calculable on both theoretical and observational grounds, would strongly suggest that dark matter cannot consist of the protons and neutrons that normally make up baryonic matter (Greene, 2004, pp. 432–435). The contribution of dark matter to the cosmic mass–energy density has been estimated by its gravitational effects. However, at around 22%, its contribution was still not enough to bring the total even close to the critical value at which theorists insist it should lie.

Fortunately, another potential contributor has recently made its appearance. Evidence from one type of supernova seems to indicate that the cosmic expansion postulated by the big bang, instead of steadily decelerating up to the present as gravity gradually slows it down, actually began to accelerate about seven billion years ago. The evidence for this unexpected finding requires an extraordinary degree of complex interpretation, but so far it seems to be solidifying.

What could be responsible for this unexpected behaviour? When Einstein encountered a somewhat similar problem (he believed the universe to be static but gravity ought to make it contract), in order to counteract gravity he inserted an essentially arbitrary cosmological term (Lambda) on the (left) side of the field equation that defines the metric space–time tensor, treating it in effect as a property of space (Einstein, 1952/1917, p. 186). When cosmic expansion was later established by Hubble, Lambda turned out not to be needed. Might one not make a similar move once again in order to account for the newly discovered acceleration of cosmic expansion? This time it would be preferable to construe the addition directly as an energy. On the (right) side of the field equation that defines the mass–energy tensor there is already a pressure term, besides the one for mass–energy. This suggests invoking a form of energy that could plausibly be treated as a negative pressure; one that could therefore account for expansion.[6]

---

[6] The derivation above recognizes the fundamental difference between Einstein's cosmological constant and the vacuum energy that could account for the cosmic acceleration (Krauss and Turner, 2004). Although they can be made mathematically equivalent, it is misleading to speak, for example, of 'moving Einstein's Lambda factor from the left to the right side of the field equation'; the pressure term was already on the right side. It is thus a little misleading to treat the newly postulated energy as, in effect, a version of Einstein's cosmological constant, labelling it in consequence as 'Lambda'. However, this usage

A vacuum energy with density that remains constant when its volume increases would do nicely.[7] As we saw earlier, there is in fact a precedent in quantum theory for a vacuum energy arising from virtual particle–antiparticle pairs. Unfortunately, the amount of this energy is somewhere around a hundred orders of magnitude too large(!): enough to blow the cosmos apart. It makes quantum theory, as it stands, impossible to reconcile with the relativity-based cosmic space–time model.[8] Still, the hope is that the long-sought theory of quantum gravity that would bring about a resolution of this troubling contradiction would, in addition, point to the needed vacuum energy density. In any event, the currently most widely accepted explanation for the cosmic expansion is a 'dark' (i.e. still mysterious!) vacuum energy.

Given that there is as yet no explanation of why the vacuum should possess an energy with the unusual

has become standard: the reigning overall cosmological theory is now generally called 'Lambda-CDM'.

[7] Because the constancy over time of the postulated vacuum energy creates some serious problems, efforts are continuing to formulate an account of vacuum energy that would make its density decrease with time (Peebles and Ratra, 2003). A different way of supplying the energy needed to power the expansion has also been suggested, under the imaginative name of 'quintessence' (Caldwell and Steinhardt, 2000). This theory postulates 'tracker' fields that can vary in space and time. Although it predicts slightly slower cosmic expansion, the data at hand are as yet insufficiently precise to distinguish 'observationally' between the two hypotheses.

[8] Although this discrepancy was pointed out by Yakov Zel'dovich as long ago as 1967, it tended to be assumed that it could somehow be cancelled out eventually by finding theoretical grounds for cancelling the vacuum energy, i.e. bringing it down to zero. Now the aim is 'not quite to zero', which is technically much more difficult.

property of negative pressure,[9] let alone one with an energy density that is both unexpectedly tiny as well as possessing a quite precise value,[10] it is fortunate that when the contribution to the cosmic energy total is computed, it turns out to be just sufficient to bring the cosmic energy density to the desired critical value, closing the worrisome gap, and thus affording precious support for the vacuum-energy hypothesis. This would make dark energy by far the main contributor to the energy density of the universe: recent estimates give it approximately 74% of the whole.

The topic of dark energy is quite obviously a work in progress. There are too many unsolved issues, and there is too much dependence on observational evidence not yet within practical reach, to allow confident conclusions to be drawn. It is one thing to say that dark energy, if it existed, could account for several key cosmic features. But it is another thing entirely to explain its origins, to find a place for it in the complex web of contemporary cosmological theory. Still, even at this early stage of what may turn out to be a very long story, it is clear that the consequences

---

[9] Although 'negative pressure' sounds counterintuitive, Roger Boscovich and others among Newton's successors speculated about the consequences of introducing gravitational systems that would be partially self-repulsive, i.e. exert 'negative pressure'.

[10] Indeed, the task of finding a theoretical ground for it is so difficult that it has led physicists such as Max Tegmark and Steven Weinberg to postulate a 'multiverse' in order to make it unnecessary to find such a ground (Carr, 2009). The precision of the required value of the vacuum energy density raises once again the issue of cosmic 'fine tuning' that has given rise to so much controversy inside and outside recent cosmology (McMullin, 2008).

for the notion of what our universe is ultimately made are likely to be far-reaching.

2.4 Conclusions

2.4.1 One

The long and complicated history of the matter concept sketched here tells of a continuing effort to find the best ways to penetrate beyond the information given us by our senses to describe how things are and how they came to be. These ways have led far from the simplicities that inspired the hopes of the original inquiry. Aristotle saw matter as, among other things, the reservoir of potentiality. That role has now become the dominant one.

The fields postulated by relativity theory at the scale of the large and of quantum theory at the level of the small are designators of potentiality, of dispositions, of 'what-would-happen-if'. Energy itself is in a real sense an expression of potentiality. It almost seems that it is to the potential, rather than the actual, that reality should be attributed at the most fundamental level. Yet can there be potentialities without the actual? The potentialities here are not indefinite: they are quantified in various ways depending on the kind of field in question. Going from Aristotelian matter to the materialism of the early modern period involved a move from an indefinite potential, con-strained eventually only in terms of quantity, to spatially extended, and indisputably actual, hard massy particles. From these particles to the 'matter' of present-day physics could be described as a move back again to potentiality, although no longer indefinite.

2.4.2 *Two*

We have seen that one can go either way with the decision as to whether to limit the scope of the term 'matter' to rest mass, or to extend it to mass–energy so as to include energies related in one way or another to mass. Would that extension alone be enough to secure for matter its traditional role as a generic term for the 'stuff' of physical reality? The present status of dark energy is too uncertain to allow a confident answer. Is dark energy mass-related? It is if the Einstein equivalence holds for it. And its crucial gravitational role would seem to imply that it does. But does that mean that it could be transformed into something possessing rest mass? We have no idea. At any rate, when cosmologists speak of the cosmic 'mass density', it is significant they make it constitute only around a third of the whole, including in it only baryonic and dark matter (see, for example, Peebles and Ratra, 2003). It does not include dark energy.

Also, the implications of the term 'vacuum' ought not to be forgotten: it presumably means that the relevant space is 'empty'. Empty of what? Presumably of rest mass. To all appearances, dark energy as a property of the vacuum would have been present in the 'vacuum' even if there had never been any rest mass at all. That would seem to suggest that it is not, in fact, mass-related. In any event, it appears rather strained to qualify it as 'matter'. If that protean term is to be used at all in the cosmic context, it is best to say that the universe consists of matter *and* energy. And if one has to have a generic term to apply to the universe as a whole, then 'energy', rather than 'mass–energy', would seem (for the moment, at least) to be the

41

proper one. The density term at the cosmic level would then simply be 'energy density'.

2.4.3 Three

What, then, of 'materialism'? Ought that label to be replaced with 'energeticism'? More important than the choice of label, what of the reductionist project that was synonymous with the older materialism? One of the most significant shifts that marked the Scientific Revolution as indeed a 'revolution' was the turn to underlying physical structure as the means of explaining the properties of macroscopic bodies. The further assumption was that the components of the explanatory structure – corpuscles, ethers, and the like – could be fully and finally described in terms of primary qualities themselves familiar from everyday experience. This form of explanation could, then, plausibly be described as 'reducing' the whole to its parts.

Gravity posed a problem, as we saw; no place could be found for it in the categories of the reductive 'mechanical philosophy' of the day. Attributing gravity to matter meant inferring from matter's behaviour in an ensemble to a potential possessed even when the material body was considered on its own. Gravitational mass and inertial mass describe what *would* happen if the body in question were to act gravitationally upon, or be acted upon by, another body. These behaviours are elicited only in the presence of a larger whole. Fields of the kind that define contemporary physics are likewise holistic in nature. They exist throughout an extended region and the field values at any point depend on the values across the field generally.

Inferring from the behaviour of a whole to a quality of its parts expressed only when they are acting as parts of that whole is just the opposite of reduction, understood as explaining the behaviour of a whole in terms of the properties of its parts when these parts are considered in isolation (McMullin, 1972). The former inference might be better described as explaining the parts in terms of the whole. The whole in such a case might still be said to be nothing more in ontological terms than a collection of its parts, but only on condition that the 'parts' are defined in equally ontological terms by the role they play in the whole. Reductionism and holism point in different ways.

2.4.4 Four

The strong forms of mind–body (or soul–body) dualism of the past posited a sharp ontological divide between the immaterial and the material, where the former often tended to be defined simply as the negation of the latter. That brisk way of approaching the distinction is no longer adequate. The boundaries of 'material' potentiality, the outer limits of the forms that can emerge in the most complex wholes that energy can sustain, are nowadays not so easily set down. Quantum entanglement and particle creation ought to warn us that further surprises are almost certainly still in store.

At the same time, it should not be supposed either that these scientific developments point to a straightforward reductionist solution of the mind–body problem that has for long proved so teasing to philosophers of mind. Herbert Feigl long ago warned against drawing this inference, attractive though it may seem to some (Feigl, 1962). The

43

main objections against the reductionist account retain their force even when the limitations of Newtonian matter are replaced by the indefiniteness of twenty-first-century energy. A non-reductive solution is still called for.

2.4.5 Five

The materialism of old appealed to the familiar as the means to explain. A (supposedly) well-defined 'matter' was therefore the starting point of inquiry. Explanation in the physics of today works in just the opposite direction. It serves, rather, as a means of discovery of the unfamiliar (McMullin, 1994). It reveals new worlds that no longer fit anywhere into intuitions schooled in the everyday. It effects an enlargement that constantly opens up to new horizons. Its 'matter' is a product in the first place of imagination, one by no means to be taken for granted. It is in need, rather, of extended and rigorous test. A 'materialism' of today, if one were to retain that label, would of necessity be one that could point only to a 'material' with true lineaments that constantly recede from view.

References

Barrow, J. and Silk, J. (1993). *The Left Hand of Creation: The Origin and End of the Expanding Universe*. New York: Oxford University Press.

Blackwell, R. (1978). Descartes' concept of matter. In *The Concept of Matter in Modern Philosophy*, ed. E. McMullin. Notre Dame, IN: University of Notre Dame Press, 59–75.

Bobik, J. (1965). Matter and individuation. In *The Concept of Matter in Greek and Medieval Philosophy*, ed. E. McMullin.

Notre Dame, IN: University of Notre Dame Press, 281–292.

Caldwell, R. R. and P. Steinhardt (2000). *Quintessence*, accessed 23 March 2010. (http://physicsworld.com/cws/article/print/402).

Carr, B., ed. (2009). *Universe or Multiverse?* Cambridge: Cambridge University Press.

Davies, P. (2006). *The Goldilocks Enigma*. London: Allen Lane.

Einstein, A. (1952/1917). Cosmological considerations on the general theory of relativity. In *The Principle of Relativity*, ed. A. Einstein et al. New York: Dover, 177–188.

Eslick, L. (1965). The material substrate in Plato. In *The Concept of Matter in Greek and Medieval Philosophy*, ed. E. McMullin. Notre Dame, IN: University of Notre Dame Press, 39–54.

Feigl, H. (1962). Matter still largely material. *Philosophy of Science*, 29: 39–46.

Greene, B. (2004). *The Fabric of the Cosmos*. New York: Random House.

Holden, T. (2006). *The Architecture of Matter: Galileo to Kant*. Oxford: Clarendon Press.

Johnson, H. J. (1973). Changing concepts of matter from antiquity to Newton. In *Dictionary of the History of Ideas*, vol. 3, ed. P. P. Wiener. New York: Scribner, 185–196.

Krauss, L. (1989). *The Search for the Fifth Essence*. New York: Basic Books.

Krauss, L. M. and Turner, M. S. (2004). A cosmic conundrum. *Scientific American*, September 2004: 71–77.

McMullin, E., ed. (1965). *The Concept of Matter in Greek and Medieval Philosophy*. Notre Dame, IN: University of Notre Dame Press.

McMullin, E. (1972). The dialectics of reduction. *Idealistic Studies*, 2: 95–115.

McMullin, E., ed. (1978a). *The Concept of Matter in Modern Philosophy*. Notre Dame, IN: University of Notre Dame Press.

McMullin, E. (1978b). *Newton on Matter and Activity*. Notre Dame, IN: University of Notre Dame Press.

McMullin, E. (1992). *The Inference That Makes Science*. Milwaukee, WI: Marquette University Press.

McMullin, E. (1994). Enlarging the known world. In *Physics and Our View of the World*, ed. J. Hilgevoord. Cambridge: Cambridge University Press, 79–113.

McMullin, E. (2002). The origins of the field concept. *Physics in Perspective*, 4: 13–39.

McMullin, E. (2008). Tuning fine-tuning. In *Fitness of the Cosmos: Biochemistry and Fine Tuning*, ed. J. Barrow et al. Cambridge: Cambridge University Press, 70–94.

Peebles, P. J. E., and Ratra, B. (2003). The cosmological constant and dark energy. *Reviews of Modern Physics*, 75: 559–606.

SLAC (2006). *Virtual Visitor Center. Theory: Real and virtual particles*, originally accessed 10 June 2006; last updated 15 June 2009 (http://www2.slac.stanford.edu/vvc/theory/virtual.html).

Thackray, A. (1968). 'Matter in a nutshell': Newton's *Opticks* and eighteenth-century chemistry. *Ambix*, 15: 29–53.

Weisheipl, J. (1965). The concept of matter in fourteenth-century science. In *The Concept of Matter in Greek and Medieval Philosophy*, ed. E. McMullin. Notre Dame, IN: University of Notre Dame Press, 147–169.

3

Unsolved dilemmas: the concept of matter in the history of philosophy and in contemporary physics

PHILIP CLAYTON

~

By the end of the modern period, a particular world view had become firmly entrenched in the public understanding. Unlike most philosophical positions, which are sharply distinguished from scientific theories, this world view was widely seen as a direct implication of science, and even as the *sine qua non* for all scientific activity. For shorthand, let's call this view "materialism."

Materialism consisted of five central theses:

(1) Matter is the fundamental constituent of the natural world.
(2) Forces act on matter.
(3) The fundamental material particles or "atoms" – together with the fundamental physical forces, whatever they turn out to be – determine the motion of all objects in nature. Thus materialism entails determinism.
(4) All more complex objects that we encounter in the natural world are aggregates of these fundamental

Information and the Nature of Reality: From Physics to Metaphysics, eds. Paul Davies and Niels Henrik Gregersen. Published by Cambridge University Press © P. Davies and N. Gregersen 2010, 2014.

particles, and their motions and behaviors can ultimately be understood in terms of the fundamental physical forces acting on them. Nothing exists that is not the product of these same particles and forces. In particular, there are no uniquely biological forces (vitalism or "entelechies"), no conscious forces (dualism), and no divine forces (what came to be known as supernaturalism). Thus materialism implied the exclusion of dualism,[1] downward causation (Bøgh Andersen et al., 2000), and divine activity.[2]

(5) Materialism is an *ontological* position, as it specifies what kinds of things do and do not exist. But it can also become a thesis concerning what may and may not count as a scientific explanation. When combined with a commitment to scientific reduction, for example, it entails that all scientific explanations should ultimately be reducible to the explanations of fundamental physics. Any other science, say biology or psychology, is incomplete until we uncover the laws that link its phenomena with physics. In its reductionist form – which historically has been its most typical form – materialism thus excludes interpretations of science that allow for "top-down" causation, also known as "strong emergence."[3] Materialists may be divided on

---

[1] The classic materialist view of consciousness is expressed in Crick (1994).

[2] The challenge to divine action is well spelled out in a series of books, *Scientific Perspectives on Divine Action*, edited by R. J. Russell, and published by the Vatican Observatory Press. See especially the summary volume in the series by R. J. Russell, N. Murphy, and W. Stoeger (2008).

[3] The distinction between "strong" emergence, which affirms real causal activity at levels of organization higher than physics, and "weak"

whether, and if so how soon, these reductions will actually be accomplished. Still, it is an entailment of materialism in most of its modern forms that an omniscient knower would be able to reduce all higher-order phenomena to the locations and momentums of fundamental particles.

In the following pages I argue that we have both philosophical and scientific reasons to doubt the adequacy of this widely accepted doctrine of materialism. In the history of Western philosophy, as we will see, it has turned out to be notoriously difficult to formulate a viable concept of matter. And physics in the twentieth century has produced weighty reasons to think that some of the core tenets of materialism were mistaken. These results, when combined with the new theories of information, complexity, and emergence summarized elsewhere in this volume, point toward alternative accounts of the natural world that deserve careful attention and critical evaluation.

3.1 The concept of matter in the history of philosophy

A strange dynamic emerges when one begins to study the history of the concept of matter in Western philosophy. It appears that, each time the greatest systematic philosophers have attempted to define it, it has receded again and again from their grasp. The very philosophers who claim to offer a resolution of the conceptual problems

emergence, which denies this, goes back to Bedau (1997, pp. 375–399). For more detail on these concepts, see Clayton (2004).

and a synthesis of opposing schools – Plato, Aristotle, Thomas Aquinas, Descartes, Leibniz, Hegel, Whitehead – repeatedly fail to supply a substantive concept of matter, leaving the reader each time merely with lack, or *privatio*: nothing instead of something. When one adds the recurring paradoxes that arise within philosophical theories of matter to the developments in physics sketched in the following section, one begins to wonder whether there is something fundamentally flawed in the idea of a world built up out of matter.

Although the description just given applies to a whole series of philosophers in the West, it fits the philosophy of Plato with particular accuracy. Plato inherited a rich tradition of natural philosophy developed during the pre-Socratic period. Numerous philosophers had developed divergent accounts of what could be the *archē*, or ultimate principle, which for many amounted to an account of the nature and properties of matter. Thus for Thales all was ultimately water; for Empedocles it was the four elements of earth, air, fire, and water; for Parmenides, the *logos*, or reason; and for Heraclitus, the principle of change itself ("you can never step into the same river twice"). Plato realized that this diversity of incompatible positions confronted philosophy with a series of dilemmas: Is everything part of a single unity, or does "the many" represent the ultimate truth? Is change real, or is it illusory? What unifies the diversity of appearances? As is well known, Plato found his solution in the doctrine of the "Forms." What is ultimately real is the *eidos*: the idea of a thing. These ideas exist in a purely intellectual realm and serve as the patterns or exemplars after which all existing things are modeled. This object is a tree because it participates in the form of treeness, and that is a just state because it participates in the form of justice.

However, Plato's theory had an unfortunate consequence, for it implied that the material world must be in some sense illusory. What is ultimately real are the forms; hence, to the extent that a form is embodied, it becomes less real. Thus, in the famous illustration of the divided line at the end of Book 6 of the *Republic* (Plato, 2000, 509d–513a), the further one descends from the realm of the forms, the less reality is possessed by the objects one encounters. They represent *doxa*, or mere opinion. The Myth of the Cave at the beginning of Book 7 likewise shows that the "knowledge" we think we have in the realm of matter is illusory. Reality only truly emerges when one ascends to the realm of the forms. In Plato's myth, this is the world above ground as seen in light of the sun, which presumably stands for the form of the Good (ibid., 514a–529a).[4] Conversely, the movement downward from intellect to matter is simultaneously a movement from knowledge to "mere opinion" and from reality to illusion.

Aristotle, a one-time student of Plato's, was disturbed by the implication that matter might just be an illusion. At first it seemed as though his philosophical system had solved the problem. For Aristotle, each existing object was in fact a "unity" of form and matter. Hence his metaphysical proposal is known as "hylomorphism," from the Greek *hylē* ("matter") and *morphē* ("form"). Matter was supposed to answer Plato's unsolved problem: *What* is it that changes or becomes? As Mary Louise Gill shows in her classic study, Aristotle's concept of matter lies "at the

---

4 See also Plato's metaphor of the sun in Book 6, 507b–509c.

intersection of [his] theory of substance and his theory of change" (Gill, 1989, p. 3).

Matter is "that from which a product is generated that is present in [the product], as the bronze of a statue and the silver of a bowl" (Aristotle, 1934, II.3, 194, 23–26). Aristotle believed that postulating this matter as a metaphysical ultimate would allow him to explain what makes this particular object what it is:

> What is the cause of the unity of the spherical and the bronze? Indeed the difficulty disappears because the one is matter, the other form. So what is the cause of this, of something in potentiality to be in actuality, except the maker, in the case of things [for which] there is generation? For there is no other cause of the sphere in potentiality being a sphere in actuality, but this was the essence for each.
>
> (Aristotle, *Metaphysics* Z, 1933, 1045^a, 25–33)

Gill comments, "According to the account of composites in Z, matter like bronze is a distinct subject to which the spherical shape belongs. Thus, the shape is in the bronze, as this form in this matter ... The problem of unity for material composites is, one must specify two distinct things – the matter and the form" (Gill, 1989, p. 142).

The problems begin to arise when one seeks to understand what this matter is actually supposed to be. Aristotle never succeeded in developing a systematic theory of matter, and additional postulates are required to make his scattered comments consistent. Freudenthal notes in his work on Aristotle's theory of material substance:

> [I]n [his] theory of matter there is no "necessitation from below": Aristotle's matter does not organize itself spontaneously into structured substances such as living beings. But, obviously,

forms emerge in matter – living beings come to be…It follows that the account of structures existing in the material world cannot be given within the framework of Aristotle's sole theory of matter, and so must involve additional explanatory postulates.   (Freudenthal, 1995, p. 2)

In fact, the problem is worse. In Aristotle's system, whenever some thing is differentiated from other things – whenever it is *this* rather than *that* – it is distinguished thanks to its form. Pure matter, then, must be purely undifferentiated stuff. Matter is the *hypokeimenon* (ὑποκείμενον), that which lies beneath (cf. the Latin *subjectum*); it is what takes on all the properties of the thing without itself having any intrinsic properties. But if it has no form and properties of its own, it cannot be directly grasped by reason. Matter as *hypokeimenon* stands closer to the idea of *khôra* ("receptacle") in Plato's *Timaeus* (1965) – the container or space in which something else takes place.[5] Matter is that unknown which, when combined with form, produces this or that specific object. But taken by itself it is completely unknown, mysterious. Matter is that which forever eludes the grasp of the philosopher. (Perhaps this embarrassing consequence of Aristotle's philosophy has something to do with the fact that, when experimental natural science started to emerge in the early modern period, it found itself forced to break free from the strictures of Aristotelian natural philosophy and to begin again on a different basis.)

Predictably, during the many centuries dominated by Platonism, the same difficulties arose that we noted above.

---

[5] This is also the sense in which Jacques Derrida uses the term in his famous little book on khôra. See Derrida (1995a).

Plato's great disciple, Augustine, faithfully passed the Platonic view of matter into the tradition of Christian philosophy, where it remained dominant in the West for the next 1000 years. Both matter and evil represented a privation of being or goodness (*privatio boni*) rather than positive principles in their own right. Even Plotinus, the great mystical philosopher who sought to synthesize Plato and Aristotle in the third century, continued the tradition of locating essential reality at, or above, the level of intellect. For him, as for the Gnostic religious philosophies of the Hellenistic period, matter was that from which one must flee in order to experience salvation or liberation – or knowledge. A similar idealist strain continued to dominate through the long history of Neo-Platonism in the West.[6]

Aristotle's old problem was repeated in the work of Thomas Aquinas in the thirteenth century. In contrast to the Platonic theologians, Aquinas sought to affirm the empirical world and to take seriously the creation of a material world by God. Following Aristotle, he viewed objects as a combination of form and matter. At first it looked as though Aquinas was able to offer a more adequate theory of matter than Aristotle because his theology allowed for the possibility that God created the matter of the universe *ex nihilo*. One might expect that the creation of the world by God would lend matter a more solid existence and assure its ontological status.

---

[6] This is masterfully demonstrated in the work of Werner Beierwaltes, for example Beierwaltes (1972, 1985) and Beierwaltes, von Balthasar, and Haas (1974).

However, Aquinas, later baptized as "the theologian" of the Catholic Church, failed to solve the conundrum of matter. Since God, the ultimate definer of Being (*esse ipsum*), is pure Spirit, not embodied in or dependent upon matter in any way, the relation of matter to God as its ultimate source remains a dilemma. How could God create something essentially different from himself? (The relation of God to evil remains equally puzzling, again suggesting the parallel that we noted in Augustine: matter ≈ evil.) The problem is reiterated in Aquinas's anthropology: the essence of the human person is the soul, which is each person's "form" or essence. If the person is to be complete, his or her soul must be reunited with the body after death. Yet the nature of this matter, which is somehow supposed to be necessary for full existence, remains unthought. To the extent that Aquinas's theology came to supply a normative framework for much of subsequent Christian theology, especially in the Roman Catholic tradition, his inadequate answer to the problem of matter continues to influence Western thinkers to the present day.

René Descartes, the so-called "father of modern philosophy" in the West, at first seemed to make progress on this ancient dilemma. In his *Meditations* of 1640 (Descartes, 1968–1969), he insisted that there are two ultimate kinds of substance: *res cogitans*, or "thought," and *res extensa*, or "matter." As the text proceeds, however, it gradually becomes clear that, although Descartes has guaranteed matter a clear ontological status, its role remains subordinate to thought. The essence of the person is the mind or consciousness, which stands in an absolute contrast to the body. Thus Descartes writes in the *Discourse*, "I knew I was a substance the whole essence of which is to think, and that for its existence there is no need of any place, nor does it depend on any material thing, so that this 'me', that is to say the soul . . . is entirely distinct from the body" (ibid., p. 101). Or, in his most pithy expression, "I exist and am not a body; otherwise, doubting of my body I should at the same time doubt myself" (ibid., p. 319).

Descartes could never solve the problem of the inter-action of mind and body because he had defined them at the outset as two diametrically opposed substances with no common ground.[7] Faced with this sort of ultimate dichotomy, all that remains is to center one's philosophi-cal system on the one or the other. Descartes, still deeply influenced by the disembodied God of Western theism, made the (for him) obvious choice and placed all value upon the side of mind, will, and rationality.

Gottfried Wilhelm Leibniz represents a particularly interesting instance. His metaphysical theories were

---

[7] Thus Julius Weinberg writes, "It can be shown that Descartes has two different arguments for the distinction of mind and body. (1) It is possible, i.e. involves no contradiction, to think that I, as thinking, exist and that nothing extended exists, and since the existence, power, and veracity of God assures me that God can bring about whatever I can conceive, it is therefore possible that I exist without a body. Hence, body and mind are really distinct. (2) The essence or attribute of nature which is thought (*cogitatio*) is logically incompatible with that of extension. Hence these attributes cannot belong to one substance but only to two" (Weinberg, 1977, p. 71). Weinberg adds, "Descartes' interest in the proof of a real distinction between mind and body is, at least, twofold. On the one hand, it forms the basis of a proof of the immortality of the soul . . . On the other hand, Cartesian dualism opens the way to a purely physical or even mechanical account of the physiology of the human body and, indeed, a purely physical account of the natural world" (Weinberg, 1977, p. 72).

highly influential, and through his disciples Wolff and Baumgarten remained dominant in European thought until the time of Kant. Leibniz was deeply intrigued by the development of mechanistic physics in the seventeenth century and contributed to its development in a major way through the invention of the differential calculus. His philosophy of infinitely divisible particles would, he believed, provide a metaphysical platform for unifying this new physics with the Western metaphysical tradition, and with Christian theology in particular. This meant, however, that Leibniz had to show how the resulting universe could be created and ruled by God, could be purposive and meaningful, and could be compatible with the perfect goodness of its omnipotent Creator. With this goal in mind, he defined the existence of individual atoms or "monads" as *purely mental* sources of activity:

The Monad, of which we will speak here, is nothing else than a simple substance, which goes to make up composites; by simple, we mean without parts ... There is nothing besides perceptions and their changes to be found in the simple substances. And it is in these alone that all the internal activities of the simple substance can consist.   (Leibniz, 1992, pp. 67, 70)

In his lengthy correspondences, Leibniz tried to work out an adequate theory of matter. Taken all together, he argues, the "simple substances" produce the behaviors in the world that physicists study. But individually, each one is as we are: a center of intellectual activity, will, and understanding. Cells and electrons may possess much less understanding than we humans do, but they are mental agents nonetheless. Further, each monad is "windowless," which means that it does not actually perceive its surrounding particles *and is not influenced by them in any way*:

In a way, then, we might properly say, although it seems strange, that a particular substance never acts upon another particular substance nor is it acted upon by it. That which happens to each one is only the consequence of its complete idea or concept, since this idea already includes all the predicates and expresses the whole universe... There is also no way of explaining how a Monad can be altered or changed in its inner being by another created thing, since there is no possibility of transposition within it, nor can we conceive of any internal movement which can be produced, directed, increased or diminished there within the substance... The Monads have no windows through which anything may come in or go out. (Leibniz, 1992, pp. 25, 68)[8]

The entire appearance of a smoothly running machine that the universe possesses is the product of a "pre-established harmony," for which God must be given the credit.

Nicholas Jolley,[9] a well-known Leibniz scholar, thinks that it is obvious that Leibniz in the end reduces what we call matter to a merely epiphenomenal property of the monads. He cites a passage from a letter Leibniz wrote to De Volder: "I do not really eliminate body, but I reduce it to what it is. For I show that corporeal mass, which is

---

[8] See Leibniz (*Discourse*, p. 23): "God produces different substances according to the different views which he has of the world, and by the intervention of God, the appropriate nature of each substance brings it about that what happens to one corresponds to what happens to all the others, without, however, their acting upon one another directly."

[9] See Jolley (1993, pp. 384–423, especially p. 399). I treat Leibniz in greater detail in Chapter 4 of *The Problem of God in Modern Thought* (Clayton, 2000).

thought to have something over and above simple substance, is not a substance, but a phenomenon resulting
from simple substances, which alone have unity and absolute reality."[10] In the same letter Leibniz writes that bodies
are just "sets of harmonized perceptions."[11]

There is some evidence to suggest that Leibniz never
found a position on matter that he was satisfied with,
and as his correspondences with De Volder and Bernoulli
progress he continually weakens the concept of matter
presupposed there. He seems most concerned to show
that the laws of nature can be preserved even under a phe-
nomenalist theory of matter. *Sometimes* Leibniz pushes his
system to its logical conclusion and ends up with idealism.
But more often he is working, as L. J. Russell argues, "to
escape the extreme interpretation of the doctrine of substance to which his metaphysical and logical speculations
of 1686 had led."[12]

Does he succeed in the end? Most commentators say no.
The conclusion of Georges Friedmann's masterful comparison between Leibniz and Spinoza is that "The philosophy of Leibniz is, at core, a monism of the spirit"; or at least
it is a philosophy "where, despite the efforts of the author,
the reality of matter and its borders with spirit are evasive

---

[10] "*Ego vero non tollo corpus, sed ad id quod est revoco, massam enim corpoream
quae aliquid praeter substantias simplices habere creditur, non substantiam
esse ostendo, sed phaenomenon resultans ex substantiis simplicibus quae solae
unitatem et absolutam realitatem habent*" (Gerhardt, 1875–1890, p. 275).
Translated in Leibniz (1989, p. 181).

[11] The process of development is nicely summarized in L. J. Russell
(1981, pp. 104–118). Bernoulli's best reconstruction is that Leibniz
makes *materia secunda* out of points endowed with forms; a material
substance is a *punctum cum forma* (Gerhardt, 1849–1863, 2:546f.).

[12] Here I follow L. J. Russell (1981, p. 118).

and fragile" (Friedmann, 1962, p. 245f). Throughout the correspondences Leibniz continues to speak as if there is matter, and hence motion and empirical perception. He does well to do so, as a thoroughgoing idealism would make it more difficult (to put it mildly) to individuate the mental substances that are the building blocks of his metaphysics. But matter is at best a by-product of the mental substances, and at worst an illusory category incompatible with what is at root an idealist system. Thus, it appears, the first major metaphysical system written after the dawn of modern physics turns out to be a form of unmitigated idealism. Matter, it seems, is merely an appearance, an illusion foisted upon us by an inaccurate comprehension of the world around us.

Nineteenth-century German philosopher Georg Wilhelm Friedrich Hegel claimed to offer the great philosophical synthesis of all knowledge and of all previous philosophies. He believed that the dichotomy between mind and matter, like all previous dichotomies, was something he could leave behind. In Hegel's writings, one does indeed find numerous attempts to incorporate the results of the natural science of his day. Unfortunately, however, in the development of Hegel's system the concept of mind or spirit (*Geist*) dominates yet again. Although one may not perceive it fully until the end of history, the force that moves all things and propels history forward is Absolute Spirit, not matter. *The Phenomenology of Spirit* (1807) chronicles the history of "Spirit coming to itself"; the history of Spirit, it turns out, provides the ultimate explanation and the ultimate moving force for all that is. If there is a material aspect of the Absolute, it remains strangely silent in Hegel's work. In the end, matter does not play any stronger role than in the work of Hegel's predecessors. As Stojanow notes correctly, "Hegel, abiding by his purely epistemological approach, examines only the ideal side, only the actuality, the pure activity, *actus purus*; he abstracts the latter from the material entelechy. According to Hegel matter is the purely passive substratum of each alteration, becoming and activity" (Stojanow, 2001).

Of course, Hegel does include a philosophy of nature as part two of his massive *Encyclopedia of the Philosophical Sciences* (1830). Some scholars have used contemporary science to attempt to vindicate Hegel's philosophy of nature.[13] But overall, I suggest, the growth of science has rendered much of Hegel's philosophy of nature obsolete.[14] Indeed, the subordination of the material moment is implicit already in his theory of self-consciousness. As he writes in the *History of Philosophy*:

The meaning . . . is not, however, that natural objects have thus themselves the power of thinking, but as they are subjectively thought by me, my thought is thus also the Notion of the thing, which therefore constitutes its absolute substance. . . . *It is only in thought that there is present a true harmony between objective and subjective*, which constitutes me.   (Hegel, 1974, pp. 149–150)

Finally, one must add the name of perhaps the greatest Western metaphysician of the twentieth century, Alfred North Whitehead. Whitehead was the first major metaphysical thinker to write his system after the breakthroughs of Einstein's theories of special and general relativity. His *Process and Reality* (1929) is meant to be an

---

[13] See, for example, Burbidge (1996).   [14] See Houlgate (1998).

empirically sensitive work, a response to scientific developments in physics and cosmology that is continually open to revision. But commentators have also recognized that Whitehead's proposals are deeply reliant on something like a Leibnizian atomism. The great Whitehead commentator, David Ray Griffin, has shown that Whitehead's system amounts to a form of "pan-experientialism" (Griffin, 2001). Every part of the physical world consists of individual moments of experience or "actual occasions." Each actual occasion receives the input of its environment as data and then synthesizes it according to its own unique moment of creativity. This metaphysic yields the startling conclusion that the fundamental constituents of the universe are not merely material; every part of the universe also includes an element of mental experience. Actual occasions at lower levels of the natural hierarchy obviously have experiences that are less complex and less rational than our own; still, they remain genuine experiences. Whitehead does speak of both a mental and a physical pole in the experience of each unit of reality (each "actual occasion").[15] But he generally places the stress on the creative process; the physical dimension concerns what is left over after the process of becoming ("concrescence") has ended. What we call materiality is always in part a by-product of earlier acts of creative synthesis by the "actual occasions" of the past.

In this brief sketch of the history of Western metaphysics, we have seen that the problem of matter remains an unsolved conundrum. Although the problem was

[15] For references to texts on the "mental" and "physical" poles, see the index to Whitehead (1978).

continually reformulated and redefined, every attempt to understand matter ends up focusing on the active principle of the intellect – that which makes understanding possible – rather than on what was to be understood, which was matter *qua* non-mental. Again, it is as if matter continually recedes from our grasp. One even wonders: Could it be that matter is in its essence *that which cannot be understood*, that which inevitably recedes from us as we approach it? Here one thinks of the notion of the "transcendental signified" in the work of the influential French philosopher, Jacques Derrida (1995b, 1998). If the parallel indeed holds, matter is another name for what Derrida called *la différance*: that which is always *different* from our formulations and which is always *deferred* into the future whenever we seek to understand it. One suspects that "matter" is being used simply as another name for the Unknown.

3.2 Matter in contemporary physics

In the opening we reviewed the fundamental tenets of materialism: reality consists of fundamental material particles; these basic particles, together with the forces that act upon them, determine the behavior of all objects in the world; all else is built up out of these constituents; and reductionism is true. In Steven Weinberg's reductionist program, for example, all causal arrows point upward from the fundamental microphysical causes, and all explanatory arrows point downward (Weinberg, 1994).

Initially, it would seem that physics offers a much more useful approach for understanding matter than does metaphysics. After all, physical science was born out of the decision to eschew the vagaries of metaphysical reflection and to work instead to "save the appearances." The goal of physics is to construct a series of hypotheses that adequately describe, explain, and predict the movements (dynamics) of the objects we observe. For several hundred years, physicists succeeded beyond all expectation at deriving principles and laws capable of explaining the behavior of objects in the physical world based on this world view.

The materialist program of research was admirably described in Book One of Thomas Hobbes' great seventeenth-century work, *Leviathan* (1651), and in John Locke's epistemological method in the *Essay Concerning Human Understanding* (1690). Hobbes began with the premise that all is "matter in motion"; the crucial task that then arises is to understand exactly *how* things move – and how the illusion that non-material things exist might have arisen out of the purely material world. This program has been called the "analytic" or "compositional" method: one identifies the basic building blocks of reality and then shows how more and more complex wholes are constructed out of these building blocks, until one had reconstructed the full world of human experience. Run the tape backwards – that is, *decompose* or deconstruct experience into smaller and smaller parts – and you arrive again at the fundamental constituents of reality. Classical examples of this research program include Locke's exclusion of secondary qualities and Hume's exclusion of enduring subjects and metaphysical causes, viz. anything beyond "constant conjunction."

The initial results of the materialist program in early modern science were astounding. The attempt to formulate fundamental laws of motion in the work of Galileo,

Kepler, and especially Newton were staggeringly successful; as Alexander Pope wrote, "Nature and nature's laws
lay hid in night; God said 'Let Newton be' and all was
light." Newton's mechanics seemed to offer support for
the metaphysical position of materialism, the view that all
things are composed of matter. His laws in the *Principia* for
the first time provided explanatory principles that could
explain the motion of all objects, from falling bodies to distant planets. The laws presupposed a primary matter on
which forces such as kinetic energy and the gravitational
force act. Newton's second law, for example, specified the
exact relationship between force and the mass of a particular object: $f = ma$. In a similar manner, he defined the
force of gravity in terms of the mass of the two attracting
bodies using the equation

$$ F = \frac{Gm_1m_2}{r^2} \tag{3.1} $$

That is, two bodies attract each other with equal and opposite forces; the magnitude of this force is proportional to
the product of the two masses and is also proportional to
the inverse square of the distance between the centers of
mass of the two bodies.

With the work of Priestley, Lavoisier and others, chemistry yielded her secrets to what appeared to be the same
method of analysis. At first blush, it appeared that biology would offer yet another example of the same principle. All one had to do, it appeared, was to dispense with
medieval metaphysical assumptions – the plenum, the a
priori hierarchy of unchanging species, and the assumption of divine purpose – and an equally physics-based biological science could emerge. This, at any rate, is how

Darwin's breakthroughs were viewed during most of the twentieth century. For a while, it was even believed that Comte's manifesto had opened the door for similar successes in the social sciences, and that we were well on our way toward realizing that goal through the work of Durkheim in sociology, Tyler in anthropology, and the early Freud in psycho-physiology.

If all composites reduce down to basic parts, and if, once given the basic parts and fundamental physical laws, one can reconstruct the compositional process up to and including the most complicated entities and behaviors of which we are aware, then all existing objects (it was assumed) must be something like mereological sums of basic units of matter, and their identity conditions must be specifiable in this fashion. The successes of the various sciences from physics to neuroscience, combined with the shared method that all scientists seem to use, were taken to be sufficient to establish this conclusion.

Of course, this widespread response left the mind–body problem unsolved, because consciousness does not appear in the equations. Nonetheless, many materialists viewed this fact as a merely temporary embarrassment. The natural sciences would eventually succeed in understanding all human thought in terms of the neural structures, chemical composition, and electrodynamics of the brain and central nervous system. When the knowledge of all things has been reduced to fundamental particles and to universal physical laws, they maintained, the victory of materialism will be complete.

Yet somewhere near the beginning of the last century, the project of materialist reduction began to run into increasing difficulties. Special and general relativity, and especially the development of quantum mechanics, represented a series of setbacks to the dreams of reductionist materialism, and perhaps a permanent end to the materialist project in anything like its classical form. (Arguably, the electrodynamics of Faraday and Maxwell, together with the science of thermodynamics, already began to suggest revisions to classical materialism, but this claim is more widely contested.) In the contemporary scientific picture of the world, the most fundamental level of analysis of the physical world is quantum physics, the study of the subatomic particles and energies of which the macrophysical world is composed. (Physicists may someday be able to demonstrate that quantum physics is a limit case of a more fundamental set of equations, such as those of string theory or M-theory, but no such demonstration has yet been produced.)

Note, however, that the physics of the quantum world bears scant resemblance to the physics that produced and justified the classical formulations of materialism. Consider the following anomalies:

(1) Physical particles such as electrons are at this scale convertible to pulsations of energy or waves. With Röntgen's discovery of the phenomenon of radioactivity, in which solid objects gradually convert themselves into radioactive waves, physicists began to realize that there is no fundamental ontological division between matter and energy. Einstein's famous equation, $E = mc^2$, probably the best-known physics equation of all time, offered a precise quantitative recipe for converting mass to energy (or energy to mass) and has been repeatedly verified by experiment since his

time. (The American use of atomic weapons on the cities of Hiroshima and Nagasaki subsequently provided the world with an unforgettable lesson on what it means to multiply m by such a large quantity as c^2.)

(2) This convertibility was given canonical status in Schrödinger's wave equation and took popular form in von Weizsäcker's principle of complementarity.[16] Complementarity in physics means that a single (mathematically well-defined) phenomenon can be described in multiple, apparently incompatible ways – for example, as both a wave and a particle – depending on the interests of the observer and the experiment she or he designs.

(3) Under the standard ("Copenhagen") interpretation of quantum physics, the world cannot be understood as ultimately determinate. Heisenberg's principle of indeterminacy means, on this view, not only that we *cannot know* the precise location and momentum of a subatomic particle, but also that the particles themselves simply *do not have* a precise location and momentum (Heisenberg, 2007/1958). But how is one to conceive matter if the physical world lacks precise location and momentum at its most fundamental level? Certainly the standard conception of matter – as involving billiard-ball-like objects that are at a certain place at a certain time and have a specific momentum at every moment of time – collapses with the discovery of indeterminacy.

(4) It is also well known that the indeterminate nature of quantum states is resolved into a precise state at the

---

[16] See von Weizsäcker (1976, 1980).

moment of measurement. The mathematics describes a superposition of possible measurements, which is resolved at the moment of measurement into a single observed state.[17] This phenomenon, known as the "collapse of the wave function," suggests that the observer plays some constitutive role in *making the physical world become what we perceive it to be at the macrophysical level* – a collection of clearly defined and locatable objects.[18] But the physical world that is constituted or constructed (to whatever extent) by subjective observers is hardly the material world conceived by classical physics!

(5) A later by-product of research in quantum physics has been quantum field theory. From the standpoint of field theory, individual subatomic particles are expressed as "localizations" of the quantum field at a particular place and time. Thus the famous French physicist Bernard d'Espagnat argues that it is no longer accurate to understand objects as objects; they should really be understood as *properties* of a field: ways in which the field is manifested at a particular place and time. To d'Espagnat, a French example comes to mind: what we used to think of as quantum particles turn out to be less like the Eiffel Tower than

---

[17] At least that's the account given in the classical or "Copenhagen" interpretation of quantum mechanics. The "decoherence" school argues that the interaction of subatomic particles with the macrophysical world can be enough by itself to resolve the superposition into a single "coherent" macrophysically observable state. See Joos (2006, pp. 53–78).

[18] The distance between this new view of the world and classical mechanics is emphasized in several of the contributions to Russell et al. (2000).

like some qualities that are in (or, that we observe in) the Eiffel Tower, such as its height, size, or shape. So we must ask: what is it that these qualities are qualities *of*? According to d'Espagnat's controversial book, *In Search of Reality* (d'Espagnat, 1983), the only possible answer is that the quantum state vector expresses properties of some deeper underlying reality. Since we know its manifestations to us – we know what it is like when measured – and since quantum physics forbids us to speak about what it is "really like" when not measured, d'Espagnat speaks of it as a "veiled reality" (d'Espagnat, 1995). His is a sort of realism at a distance: we cannot say that reality is "just this way or that," since our observations and what we observe are intertwined; and yet we *can* say that the-world-as-observed is a manifestation of the real; reality really takes this or that form *in* our observations. Unfortunately, for traditional theories of matter, however, this "veiled reality" can be neither mental nor material, insofar as it precedes the mind–matter distinction altogether.[19]

(6) Finally, more recent work on nonlocality further undercuts classically materialist views of matter. In a series of experiments, initially designed by Alain Aspect to demonstrate the violation of Bell's inequalities, two linked photons are fired in opposite directions. A measurement of the spin made on one particle instantly resolves (creates?) the plane of spin of the other one. Yet, given the distances, which in the

[19] See Chapter 19, "The 'ground of things'" in the last great work of d'Espagnat (2006, pp. 429–464).

experiments now exceed 10 km, no message *could* be sent between the two particles except at a velocity that exceeds the speed of light, which is impossible. Some argue that the experiments demonstrate the possibility of superluminal action at a distance, whereas others maintain that two particles separated by great distances can still act as one object (Grib and Rodrigues, Jr, 1999; Maudlin, 2002). In either case, the experiments force upon us a view of the physical world that lies well outside any common-sense conception of matter.

Over-eager authors have jumped on these results, attempting to argue that they spell the end of physics or promise the final convergence of science and religion. One frequently finds titles such as: *Atoms, Snowflakes and God: The Convergence of Science and Religion*; *The Science of God: The Convergence of Scientific and Biblical Wisdom*; *The Tao of Physics*; and so forth (Capra, 1984; Hitchcock, 1982; Schroeder, 1998). Such conclusions are unjustified. Quantum physics is not a threat to physics but one of its most impressive successes in the last century. It is, however, a threat *to a particular understanding of physics*, for it is ultimately incompatible with the world view of materialism that dominated much of the physics of the modern period (Stapp, 2004).[20] It is perhaps not an overstatement to say that the developments in physics briefly summarized here provide a powerful empirical refutation of that materialist world view.

[20] See also his contribution to this volume: Chapter 6.

3.3 Toward a new scientific world view

The two preceding sections reveal some of the conundrums that face any theory of matter today. Careful empirical study of the natural world has replaced classical concepts of matter with the strange, strange view of the world offered by contemporary physics. Physicists began with a solid concept of matter and with the world of everyday experience. But as they tested this concept in light of the scientific method and, in particular, the demands of the mathematical formalisms, they were led to results that cast into question all previous conceptions of matter. One has the sense that, at the end of the day, the speculation of the philosophers and the data from the scientists are pointing in the same surprising direction. At the root of all physical reality is not "primary matter" or little atoms of "stuff." Relativity theory in cosmology and the complementarity thesis in quantum physics suggest that the basic reality is some sort of hybrid "matter–energy." Quantum field theory and string theory (if it survives as a physical theory, which now seems unlikely) suggest the even more radical idea that this reality is more energy-like than matter-like. Either result is sufficient to falsify materialism in anything like the form that dominated the first 300 years of modern science.

So what do we conclude? I suggest that the lesson is twofold. On the one hand, those thinkers are misguided who seek to dispense with the notion of matter altogether. The conundrums are not resolved by turning one's back on the mysterious nature of objects and particles in physics. Idealists who abandon the scientific study of the physical world in favor of mentalism or spiritualism

72

"solve" the dilemma by ignoring the very fields in which it can be most fruitfully studied. After all, we are surrounded by physical objects. The things we touch and manipulate are not mere figments of our imagination; unlike our ideas, their "brute existence" frequently resists our will and wishes. Since there is no evidence that all objects are thinking, perceiving beings like us, we *should* take them to be different from ourselves, physical objects without mentality (in contrast to panpsychism). The commitment to do so launches one into the research program of contemporary science.

On the other hand, as we pursue the project of science, we discover that no simple concept of matter is adequate to the results of physics – or, as we also saw, to the demands of systematic philosophy. No "primary matter" serves as the basic stuff out of which all else is composed. Instead, the deeper one pursues the explanations, the more non-materiality reveals itself in (or behind) the solid objects around us. Beginning with the oft-repeated observation that the solid object one touches is in fact composed mostly of space, one finds oneself confronted with as strange a world in physics as one will ever meet in the history of philosophy.

What is necessary, I suggest, is that we pursue this path of natural science as far as it can take us. No over-quick leaps into metaphysics will help; all such shortcuts leave one poorer in the end. But refusing to acknowledge the complex philosophical issues raised by today's science equally impoverishes human understanding. Only a partnership of scientists and philosophers will make it possible to formulate an adequate post-materialist theory of the natural world.

The move beyond materialism may have started with physicists; it is to them that we owe many of the revolutionary new concepts, including the radically new notions of information explored elsewhere in this volume. Physics suggests theories of reality in which *information* takes over the roles that matter once played, as in John Wheeler's slogan "it from bit" and Anton Zeilinger's recent extension of Wheeler's work into experimental quantum physics.[21] Physics may also suggest an "entanglement of matter and meaning."[22] But in recent years the baton has passed to the biological sciences, where new insights into the nature of information are now receiving empirical support. One thinks in particular of the studies of "top-down constraints" in systems biology,[23] biosemiotics, and form-based (morphological) theories of causality (see T. Deacon, in Chapter 8 of this volume). Taken together, I suggest, these new lines of inquiry are putting the final nails in the coffin of the materialist world view once touted as science's crowning glory.[24]

[21] See Zeilinger (2004, pp. 201–220).
[22] See Barad (2007). Barad is not an idealist, but she defends a form of realism that she calls "agential realism." There is a similar entanglement of the objective and subjective world in the work of quantum theorists von Neumann and Wigner, among others; see again Stapp (2004).
[23] See Chandler, Mason, and van de Vijver (2000): in particular, Josslyn (2000) and Lemke (2000); see also Palsson (2006).
[24] An ancestor of this chapter was presented in January 2003 at Bangalore University, India, as part of a conference on "The Concept of Matter in Indian Philosophical Schools and the New Physics: Understanding Knowledge Systems"; that material appeared as "The Concept of Matter in Traditional Western Philosophy and in Contemporary Physics: The Unsolved Dilemma," in Ananthamurthy et al. (2005, pp. 163–177). I am grateful to critical comments from Niels Henrik Gregersen and Mary Ann Meyers, and to Ashley Riordan for research assistance during the rewriting process.

References

Ananthamurthy, S. et al., eds. (2005). *Landscape of Matter: Conference Proceedings on the Concept of Matter*. Bangalore, India: Bangalore University Prasaranga Press.

Aristotle (1933). *Metaphysics*, trans. H. Tredennick, vol. XVII–XVIII, Loeb Classical Library Series. Cambridge, MA: Harvard University Press.

Aristotle (1934). *Physics*, trans. P. H. Wicksteed and F. M. Cornford, vol. IV–V, Loeb Classical Library Series. Cambridge, MA: Harvard University Press.

Barad, K. (2007). *Meeting the Universe Halfway: Quantum Physics and the Entanglement of Matter and Meaning*. Durham, NC: Duke University Press.

Bedau, M. (1997). Weak emergence. *Philosophical Perspectives*, vol. 11: *Mind, Causation, and World*. Atascadero, CA: Ridgeview.

Beierwaltes, W. (1972). *Platonismus und Idealismus*. Frankfurt am Main: V. Klostermann.

Beierwaltes, W. (1985). *Denken des Einen: Studien zur neuplatonischen Philosophie und ihrer Wirkungsgeschichte*. Frankfurt am Main: V. Klostermann.

Beierwaltes, W., von Balthasar, H. Urs, and Haas, A. M. (1974). *Grundfragen der Mystik*. Einsiedeln: Johannes Verlag.

Burbidge, J. W. (1996). *Real Process: How Logic and Chemistry Combine in Hegel's Philosophy of Nature*. Toronto: University of Toronto Press.

Bøgh Andersen, P. et al., eds. (2000). *Downward Causation: Minds, Bodies and Matter*. Aarhus and Oakville, CT: Aarhus University Press.

Capra, F. (1984). *The Tao of Physics: An Exploration of the Parallels between Modern Physics and Eastern Mysticism*. New York: Bantam Books.

Chandler, J., Mason, G., and van de Vijver, G., eds. (2000). *Closure: Emergent Organizations and Their Dynamics*, Annals of the New York Academy of Science Series, vol. 901. New York: New York Academy of Sciences.

Clayton, P. (2000). *The Problem of God in Modern Thought*. Grand Rapids: Eerdmans.

Clayton, P. (2004). *Mind and Emergence*. Oxford: Oxford University Press.

Crick, F. (1994). *The Astonishing Hypothesis: The Scientific Search for the Soul*. New York: Scribner.

Descartes, R. (1968–1969). *The Philosophical Works of Descartes*, trans. E. S. Haldane and G. R. T. Ross. Cambridge: Cambridge University Press.

d'Espagnat, B. (1983). *In Search of Reality*. New York: Springer Verlag.

d'Espagnat, B. (1995). *Veiled Reality: An Analysis of Present-Day Quantum Mechanical Concepts*. Reading, MA: Addison-Wesley.

d'Espagnat, B. (2006). *On Physics and Philosophy*. Princeton, NJ: Princeton University Press.

Derrida, J. (1995a). *On the Name*, ed. T. Dutoit. Stanford: Stanford University Press.

Derrida, J. (1995b). *The Gift of Death*, trans. D. Wills. Chicago, IL: University of Chicago Press.

Derrida, J. (1998/1967). *Of Grammatology*, trans. G. C. Spivak. Baltimore, MD: Johns Hopkins University Press.

Freudenthal, G. B. (1995). *Aristotle's Theory of Material Substance: Heat and Pneuma, Form and Soul*, Oxford: Clarendon Press.

Friedmann, G. (1962). *Leibniz et Spinoza*, 2nd ed. Paris: Gallimard.

Gerhardt, C. I. (1849–1863). *Die mathematischen Schriften von G. W. Leibniz*. Berlin: Weidmann.

Gerhardt, C. I. (1875–1890). *Die philosophischen Schriften von G. W. Leibniz*, 7 vols. Berlin: Weidmann. Excerpts available in English in Leibniz, G. W., *Philosophical Essays*, trans. R. Ariew and D. Garber (1989). Indianapolis: Hackett.

Gill, M. L. (1989). *Aristotle on Substance: The Paradox of Unity*. Princeton: Princeton University Press.

Grib, A. A., and Rodrigues, Jr. W. A. (1999). *Nonlocality in Quantum Physics*. New York: Kluwer Academic.

Griffin, D. R. (2001). *Reenchantment Without Supernaturalism: A Process Philosophy of Religion*. Ithaca, NY: Cornell University Press.

Hegel, G. F. W. (1974). *Hegel's Lectures on The History of Philosophy*, trans. E. S. Haldane and F. H. Simson. New York: The Humanities Press.

Heisenberg, W. (2007/1958). *Physics and Philosophy: The Revolution in Modern Science*. New York: Harper Perennial.

Hitchcock, J. L. (1982). *Atoms, Snowflakes and God: The Convergence of Science and Religion*. San Francisco: Alchemy Books.

Houlgate, S., ed. (1998). *Hegel and the Philosophy of Nature*. Albany, NY: State University of New York Press.

Jolley, N. (1993). Leibniz: Truth, knowledge and metaphysics. In *The Renaissance and Seventeenth-century Rationalism*, ed. G. H. R. Parkinson. Routledge History of Philosophy vol. 4, London: Routledge, 353–388.

Joos, E. (2006). The emergence of classicality from quantum theory. In *The Re-emergence of Emergence: The Emergentist Hypothesis from Science to Religion*, eds. P. Clayton and P. Davies. Oxford: Oxford University Press, 53–78.

Josslyn, C. (2000). Levels of control and closure in complex semiotic systems. In *Closure: Emergent Organizations and Their Dynamics*, Annals of the New York Academy of Science Series, vol. 901, eds. J. Chandler, G. Mason, and G. van de Vijver. New York: New York Academy of Sciences, 67–74.

Leibniz, G. W. (1989). *Philosophical Essays*, trans. R. Ariew and D. Garber. Indianapolis: Hackett.

Leibniz, G. W. (1992). *Discourse on Metaphysics and the Monadology*, trans. G. R. Montgomery. Buffalo, NY: Prometheus Books.

Lemke, J. (2000). Opening up closure: Semiotics across scales. In *Closure: Emergent Organizations and Their Dynamics*, Annals of the New York Academy of Science Series, vol. 901, eds. J. Chandler, G. Mason, and G. van de Vijver. New York: New York Academy of Sciences, 101–111.

Maudlin, T. (2002). *Quantum Non-Locality and Relativity: Metaphysical Intimations of Modern Physics*. Malden, MA: Blackwell Publishers.

Palsson, B. Ø. (2006). *Systems Biology: Properties of Reconstructed Networks*. Cambridge: Cambridge University Press.

Plato (1965). *Timaeus*, trans. H. D. P. Lee. Harmondsworth: Penguin.

Plato (2000). *Republic*, ed. G. R. F. Ferrari, trans. T. Griffith. Cambridge: Cambridge University Press.

Russell, L. J. (1981). The correspondence between Leibniz and De Volder. In *Leibniz: Metaphysics and Philosophy of Science*, ed. R. S. Woolhouse. Oxford: Oxford University Press.

Russell, R. J., et al., eds. (2000). *Quantum Mechanics: Scientific Perspectives on Divine Action*. Vatican: Vatican Observatory Press.

Russell, R. J., et al., eds. (2008). *Scientific Perspectives on Divine Action. Twenty Years of Challenge and Progress*. Vatican: Vatican Observatory Press.

Schroeder, G. (1998). *The Science of God: The Convergence of Scientific and Biblical Wisdom*. New York: Broadway Books.

Stapp, H. P. (2004). *Mind, Matter, and Quantum Mechanics*. Berlin: Springer.

Stojanow, J. (2001). On the Absolute Rational Will: II On the Absolute Material Entelechy, accessed June 23, 2008

(www.jgora.dialog.net.pl/OnTheAbsoluteRationalWill/
OnTheAbsoluteEntelechy.htm#14).

von Weizsäcker, C. F. (1976). *Zum Weltbild der Physik*, 12th ed. Stuttgart: S. Hirzel; trans. in Weizsäcker, *The World View of Physics*, trans. by Marjorie Grene (1952). Chicago, IL: University of Chicago Press.

von Weizsäcker, C. F. (1980). *The Unity of Nature*, trans. F. J. Zucker. New York: Farrar Straus Giroux.

Weinberg, J. R. (1977). *Ockham, Descartes, and Hume: Self-Knowledge, Substance, and Causality*. Madison, WI: The University of Wisconsin Press.

Weinberg, S. (1994). *Dreams of a Final Theory: The Scientists Search for the Ultimate Laws of Nature*. New York: Vintage Books.

Whitehead, A. N. (1978). *Process and Reality*, corrected ed, eds. D. R. Griffin and D. Sherburne. New York: Macmillan.

Zeilinger, A. (2004). Why the quantum? "It" from "bit"? A participatory universe? Three far-reaching challenges from John Archibald Wheeler and their relation to experiment. In *Science and Ultimate Reality: Quantum Theory, Cosmology and Complexity*, eds. J. Barrow, P. Davies and C. Harper, Jr., Cambridge: Cambridge University Press, 201–220.

PART II
PHYSICS

~

4

Universe from bit

PAUL DAVIES

~

"I refute it thus!" Samuel Johnson famously dismissed
Bishop George Berkeley's argument for the unreality of
matter by kicking a large stone (Boswell, 1823). In the light
of modern physics, however, Johnson's simple reasoning
evaporates. Apparently solid matter is revealed, on closer
inspection, to be almost all empty space, and the particles of which matter is composed are themselves ghostly
patterns of quantum energy, mere excitations of invisible
quantum fields, or possibly vibrating loops of string living in a ten-dimensional space–time (Greene, 1999). The
history of physics is one of successive abstractions from
daily experience and common sense, into a counterintuitive realm of mathematical forms and relationships, with
a link to the stark sense data of human observation that is
long and often tortuous. Yet at the end of the day, science
is empirical, and our finest theories must be grounded,
somehow, "in reality." But where is reality? Is it in acts
of observation of the world made by human and possibly non-human observers? In records stored in computer
or laboratory notebooks? In some objective world "out
there"? Or in a more abstract location?

Information and the Nature of Reality: From Physics to Metaphysics, eds. Paul Davies
and Niels Henrik Gregersen. Published by Cambridge University Press
© P. Davies and N. Gregersen 2010, 2014.

4.1 The ground of reality

When a physicist performs an experiment, he or she interrogates nature and receives a response that, ultimately, is in the form of discrete bits of information (think of "yes" or "no" binary answers to specific questions), the discreteness implied by the underlying quantum nature of the universe (Zeilinger, 2004). Does reality then lie in the string of bits that come back from the set of all observations and experiments – a dry sequence of ones and zeros? Do these observations merely *transfer* really-existing bits of information from an external world reality to the minds of observers, or are the bits of information *created* by the very act of observation/experiment? And – the question to which this entire discussion is directed – are bits of "classical" information the only sort of information that count in the reality game, or does an altogether different form of information underpin reality? In short, where is the ontological ground on which our impression of a really-existing universe rests?

In the well-known parable of the tower of turtles, the search for the ultimate source of existence seems to lead to an infinite regress. Terminating the tower in a "levitating superturtle" requires either a leap of faith – accepting the bottom level as an unexplained brute fact – or some mental gymnastics, such as positing a necessary being, the non-existence of which is a logical impossibility. Classical Christian theology opted for the latter, with God cast in the role of that necessary being, upholding a contingent universe. Unfortunately, the concept of a necessary being is fraught with philosophical and theological difficulties, not least of which is the fact that such a being does not bear any obvious resemblance to traditional notions of God (Ward, 1982). Nor is it clear that a necessary being is necessarily unique (there could be many necessary gods), or necessarily good, or able to create a universe (or set of universes) that is not itself already necessary (thus rendering the underpinning redundant). But if the universe is contingent, another problem arises: can a necessary being's nature, and hence choices, be contingent? In other words, can a necessary being *freely* choose to create something? (As opposed to necessarily making such-and-such a universe.) As a result of this philosophical quagmire, most theologians have abandoned the idea that God exists necessarily.

Science evaded all these complications by resting content to accept the physical universe itself, at each instant of time, as the basement level of reality, without the need for a god (necessary or otherwise) to underpin it. The latter view was well exemplified by British philosopher Bertrand Russell in a BBC radio debate with Fr Frederick Copleston (Russell, 1957). Russell expressed it bluntly: "I should say that the universe is just there, and that's all."

Sometime during the twentieth century, a major transition was made. The theory of relativity undermined the notion of absolute time and the shared reality of the state of the entire universe at each instant. Quantum mechanics then demolished the concept of an external state of reality in which all meaningful physical variables could be assigned well-defined values at all times. So a subtle shift occurred, at least among theoretical physicists, in which the ground of reality first became transferred to the laws of physics themselves, and then to their mathematical surrogates, such as Lagrangians, Hilbert spaces, etc. The logical conclusion of going down that path is to treat the physical universe as if it simply *is* mathematics. Many of my theoretical physicist colleagues do indeed regard ultimate reality as vested in the subset of mathematics that describes physical law. For them, (this subset of) mathematics is the ground of being. When, three centuries earlier, Galileo had proclaimed, "The great book of Nature can be read only by those who know the language in which it was written, and this language is mathematics" (Drake, 1957), he supposed that the mathematical laws were grounded in a deeper level – a level guaranteed and upheld by God. But today, the mathematical laws of physics are regarded by most scientists as free-floating – the levitating superturtle to which I referred above.

At this point, physics encounters its own conundrum of necessity versus contingency, as famously captured by Einstein's informal remark about whether God had any choice in his creation. What he meant by this was, could the laws of physics have been otherwise (that is, different mathematical relationships), or do they *have* to be as they are, of necessity? The problem of course is that if the laws could have been different, one can ask why they are as they are, and – loosely speaking – where these particular laws have "come from." To use a metaphor, it is as if mathematics is a wonderful warehouse richly stocked with forms and relationships, and Mother Nature passes through with a shopping trolley, judiciously selecting a handy differential equation here and an attractive symmetry group there, to use as laws for a physical universe.

The problem of the origin of the laws of physics is an acute one for physicists. Einstein's suggestion that they may turn out necessarily to possess the form that they do has little support. It is sometimes said that a truly unified theory of physics might be so tightly constrained logically that its mathematical formulation is unique. But this claim is readily refuted. It is easy to construct artificial universe models, albeit impoverished ones bearing only a superficial resemblance to the real thing, which are nevertheless mathematically and logically self-consistent. For example, many papers are written in which four space–time dimensions are replaced by two, for ease of calculation. These simplified "universes" represent possible realities, but not the "actual" reality (Davies, 2006).

Given that the universe could be otherwise, in vastly many different ways, what is it that determines the way the universe actually is? Expressed differently, given the apparently limitless number of entities that can exist, who or what gets to decide what *actually* exists? The universe contains certain things: stars, planets, atoms, living organisms . . . Why do *those* things exist rather than others? Why not pulsating green jelly, or interwoven chains, or fractal hyperspheres? The same issue arises for the laws of physics. Why does gravity obey an inverse square law rather than an inverse cubed law? Why are there two varieties of electric charge rather than four, and three "flavors" of neutrino rather than seven? Even if we had a unified theory that connected all these facts, we would still be left with the puzzle of why *that* theory is "the chosen one." Stephen Hawking has expressed this imponderable more eloquently: "What is it that breathes fire into the equations and makes a universe for them to describe?" (Hawking, 1988). Who, or what, promotes the "merely possible" to the "actually existing"?

There are two circumstances in which the problem of the "fire-breathing actualizer" – the mechanism to dignify a subset of the possible with the status of becoming "real" – is circumvented. The first circumstance is that *nothing* exists. However, we can rule that out on the basis of observation. The second is that *everything* exists, that is, everything that *can* exist *does* exist. Then no procedure is needed to select the actually existing things and separate them from the infinite set of the merely-possible-but-in-fact-non-existent things. Is this credible? Well, we cannot observe everything, and absence of evidence is not the same as evidence of absence. We cannot be sure that some particular thing we might imagine does not exist *somewhere*, perhaps beyond the reach of our most powerful instruments, or in some parallel universe.

Max Tegmark has proposed that, indeed, everything that can exist does exist, somewhere within an infinite stack of parallel worlds. "If the universe is inherently mathematical, then why was only one of the many mathematical structures singled out to describe a universe?" he challenges. "A fundamental asymmetry appears to be built into the heart of reality" (Tegmark, 2003). Tegmark's suggestion is one of many so-called multiverse models, according to which the universe we observe is but an infinitesimal fragment amid a vast, possibly infinite, ensemble of universes. In most variants of this theory, the laws of physics differ from one universe to another. That is, the laws are not absolute and universal, but more like "local by-laws" (Rees, 2001).

A knee-jerk reaction to Tegmark's version of the multiverse is that it flagrantly violates Occam's razor. But

Tegmark points out that everything can actually be simpler than something. That is, the whole can often be defined more economically than any of its parts. (The set of all integers, for example, is easily described, whereas a subset of integers consisting of, say, prime numbers, selected or not by a random coin toss, is not.) However, the notion of "everything" runs into formal conceptual problems when infinite sets are involved, and Tegmark's proposal is very ill defined, perhaps to the point of meaninglessness. In any case, very few scientists or philosophers would subscribe to Tegmark's extreme view. Even those who believe in some sort of multiverse usually stop short of supposing that literally *everything* exists.

The orthodox position seems to be that the actually-existing (as opposed to possible but non-existent) laws should simply be accepted as a brute fact, with no deeper explanation at all. Sean Carroll has expressed support for this position, in addressing the question, why those laws of physics? "That's just how things are," replies Carroll. "There is a chain of explanations concerning things that happen in the universe, which ultimately reaches to the fundamental laws of nature and stops" (Carroll, 2007). In other words, the laws of physics are "off limits" to science. We must just accept them as "given" and get on with the job of applying them.

4.2 Hidden assumptions about the laws of physics

The orthodox view of the nature of the laws of physics contains a long list of tacitly assumed properties. The laws are regarded, for example, as immutable, eternal, infinitely precise mathematical relationships that transcend the physical universe, and were imprinted on it at the moment of its birth from "outside," like a maker's mark, and have remained unchanging ever since – "cast in tablets of stone from everlasting to everlasting" was the poetic way that Wheeler put it (Wheeler, 1989). In addition, it is assumed that the physical world is affected by the laws, but the laws are completely impervious to what happens in the universe. No matter how extreme a physical state may be in terms of energy or violence, the laws change not a jot. It is not hard to discover where this picture of physical laws comes from: it is inherited directly from monotheism, which asserts that a rational being designed the universe according to a set of perfect laws. And the asymmetry between immutable laws and contingent states mirrors the asymmetry between God and nature: the universe depends utterly on God for its existence, whereas God's existence does not depend on the universe.

Historians of science are well aware that Newton and his contemporaries believed that in doing science they were uncovering the divine plan for the universe in the form of its underlying mathematical order. This was explicitly stated by René Descartes:

[I]t is God who has established the laws of nature, as a King establishes laws in his kingdom ... You will be told that if God has established these truths, he could also change them as a King changes his laws. To which it must be replied: yes, if his will can change. But I understand them as eternal and immutable. And I judge the same of God.   (Descartes, 1630)

The same conception was expressed by Spinoza:

Now, as nothing is necessarily true save only by Divine decree, it is plain that the universal laws of nature are decrees of God following from the necessity and perfection of the Divine nature...nature, therefore, always observes laws and rules which involves eternal necessity and truth, although they may not all be known to us, and therefore she keeps a fixed and immutable order.   (de Spinoza, 1670)

Clearly, then, the orthodox concept of laws of physics derives directly from theology. It is remarkable that this view has remained largely unchallenged after 300 years of secular science. Indeed, the "theological model" of the laws of physics is so ingrained in scientific thinking that it is taken for granted. The hidden assumptions behind the concept of physical laws, and their theological provenance, are simply ignored by almost all except historians of science and theologians. From the scientific standpoint, however, this uncritical acceptance of the theological model of laws leaves a lot to be desired. For a start, how do we know the laws are immutable and unchanging? Time-dependent laws have been considered occasionally (see, for example, Smolin, 2008), as well as observational tests carried out to look for evidence that some of the so-called fundamental constants of physics may in fact have changed slowly over cosmological time scales (Barrow, 2002). Particle physics suggests that the laws we observe today may actually be only effective laws, valid at relatively low energy, emergent from the big bang as the universe cooled from Planck temperatures. String theory suggests a mathematical landscape of different low-energy laws, with the possibility of different regimes in different cosmic patches, or universes – a variant on the multiverse theory (Susskind, 2005).

But even in these examples, there are fixed higher-level meta-laws that determine the pattern of lawfulness (Davies, 2006). Thus in the popular variant of the multiverse theory, called eternal inflation, there are many big bangs scattered through space and time, each "nucleating" via quantum tunneling, and thereby giving birth to a universe. As a universe cools from the violence of its origin, it inherits a set of laws, perhaps to some extent randomly (that is, as frozen accidents). To make this model work, there has to be a universe-generating mechanism operating in the overall multiverse (and in the case cited, it is based on quantum field theory and general relativity) and a set of general laws (like a string theory Lagrangian) from which a lucky dip of low-energy effective laws within each universe is available. Clearly this meta-law structure of the multiverse merely shifts the problem of the origin of the laws up a level.

Another strong influence on the orthodox concept of physical law is Platonism. Plato located numbers and geometrical structures in an abstract realm of ideal forms. This Platonic heaven contains, for example, perfect circles – as opposed to the circles we encounter in the real world, which are always flawed approximations of the ideal. Many mathematicians are Platonists, believing that mathematical objects have real existence, even though they are not situated in the physical universe. Theoretical physicists are steeped in the Platonic tradition, so they also find it natural to locate the mathematical laws of physics in a Platonic realm. The fusion of Platonism and monotheism created the powerful orthodox scientific concept of the laws of physics as ideal, perfect, infinitely precise, immutable, eternal, state-immune, unchanging mathematical forms and relationships that transcend the physical universe and reside in an abstract Platonic heaven beyond space and time.

It seems to me that after three centuries we should consider the possibility that the classical theological/Platonic model of laws is an idealization with little experimental or observational justification. Which leads naturally to the question: can we have a *theory* of laws? Instead of accepting the laws of physics as a levitating superturtle at the bottom of the stack – an unexplained brute fact – might we push beyond at least one step, and try to account for why the laws are as they are, to show that there are *reasons* for why they have the form that they do? To think creatively about this, it is necessary to jettison all the above-listed hidden assumptions. For example, we must allow that the asymmetry between laws and states may be incorrect, and reflect on what the consequences might be if the laws depend (at least to some extent) on what happens in the universe: that is, to the actual physical states. Might laws and states co-evolve, in such a way that "our world" is some sort of attractor in the product space of laws and states?

To illustrate a possible agenda along these lines, I want to concentrate on one aspect of the standard theological model of laws that is most vulnerable to falsification: namely, the assumption of infinite precision (Davies, 2006). The laws of physics are normally cast as differential equations, which embed the concepts of real numbers, and of infinite and infinitesimal quantities, as well as continuity of physical variables, such as those of space

93

and time. This assumption extends even to string theory, where the link with the world of space, time, and matter is long and tenuous in the extreme. As any experiment or observation can be conducted to finite accuracy only, to assume infinitely precise laws is obviously a wholly unjustified extrapolation – a leap of faith. To the extent that it may be a technical convenience, that is all right. But as I shall show, there are circumstances where the extrapolation may lead us astray in a testable manner.

To focus the issue, consider Laplace's famous statement about a computational demon. Laplace pointed out that the states of a closed deterministic system, such as a finite collection of particles subject to the laws of Newtonian mechanics, are completely fixed once the initial conditions are specified:

We may regard the present state of the universe as the effect of its past and the cause of its future. An intellect which at any given moment knew all of the forces that animate nature and the mutual positions of the beings that compose it, if this intellect were vast enough to submit the data to analysis, could condense into a single formula the movement of the greatest bodies of the universe and that of the lightest atom; for such an intellect nothing could be uncertain and the future just like the past would be present before its eyes.   (Laplace, 1825)

If Laplace's argument is taken seriously, then everything that happens in the universe, including Laplace's decision to write the above words and my decision to write this book chapter, are preordained. The necessary information is already contained in the state of the universe at any previous time. Laplace's statement represents the pinnacle of Newtonian clockwork mechanics, with its embedded assumption of infinitely precise theological laws – I would say the pinnacle of absurdity. It is the starting point for my challenge to the orthodox concept of physical law.[1]

4.3 It from bit

The basis of the challenge, which builds on the work of John Wheeler (1979, 1983, 1989, 1994) and Rolf Landauer (1967, 1986), sprang originally from the theory of information and computation. The traditional relationship between mathematics, physics, and information may be expressed symbolically as follows:

Mathematics → Physics → Information

According to this orthodox view, mathematical relationships are the most basic aspects of existence. The physical world is an expression of a subset of mathematical relationships, whereas information is a secondary, or derived, concept that characterizes certain specific states of matter (such as a switch being either on or off, or an electron spin being up or down). However, an alternative view is gaining in popularity: a view in which *information* is regarded as the primary entity from which physical reality is built. It is popular among scientists and mathematicians who work on the foundations of computing, and physicists who work in the theory of quantum computation.

---

[1] Although orthodox Newtonian mechanics assumes infinitely precise laws, Newton himself was more circumspect. He considered that the solar system might require an occasional divine prod to maintain its stability, a suggestion that incurred the derision of some of his contemporaries. Later Laplace would famously remark to Napoleon that he "had no need of this [divine prodding] hypothesis."

Importantly, it is not merely a technical change in perspective, but represents a radical shift in world view, well captured by Wheeler's pithy slogan "It from bit" (Wheeler and Ford, 1998). The variant I wish to explore here is to place *information* at the base of the explanatory scheme, thus:

Information → Laws of physics → Matter

After all, the laws of physics *are* informational statements: they tell us something about the way the physical world operates. This shift in perspective requires a shift in the foundational question I posed concerning the origin of the laws of physics; we may now ask about the origin and nature of the *information content* of the universe, and I refer the reader to Seth Lloyd's essay in Chapter 5 of this volume for one perspective on that question. Here I wish to address a more basic aspect of the problem, which is whether the information content of the universe is finite or infinite.

In the standard model of cosmology, in which there is a single universe that began with a big bang (representing the origin of space and time), the universe contains a finite amount of information. To see why, first note that the universe began 13.7 billion years ago, according to the latest astronomical evidence. The region of space accessible to our observations is defined by the maximum distance that light has traveled since the big bang: namely, 13.7 billion light-years. Because the speed of light is a fundamental limit, no information can travel faster than light, so the volume of space delimited by the reach of light defines a sort of horizon in space beyond which we cannot see, or be influenced by in terms of causal physical effects. Expressed differently, we cannot access any information beyond the horizon at this time. The horizon does expand with time t (like t^2), so that, in the future, the causally connected region of our universe will contain more information. In the past, it contained less. The technical term for the light horizon is "particle horizon," because it separates particles of matter we can see (in principle) from those we cannot see because there has not yet been enough time since the cosmic origin for the light from them to reach us on Earth. It is likely that there is another type of horizon, technically termed an "event horizon." It arises because the rate of expansion of the universe seems to be accelerating, implying (very crudely speaking) that some galaxies we now see flying away from us are speeding up, and will eventually recede so fast that their light will never again reach us. They will disappear across the event horizon for good. At some stage in the next few billion years, the event horizon effects will come to dominate over the particle horizon effects. By an odd coincidence, the radii of the particle and event horizons are roughly the same at the current epoch, and given the incompletely formulated nature of what I shall propose, either or both horizons may be regarded as the basis of the discussion (so I simply use the generic word "horizon" from now on).

A well-defined question is: how much information is there within the volume of space limited by the horizon? Information is quantified in bits, or binary digits, exemplified by a coin toss. The coin is either heads or tails, and determining which amounts to acquiring precisely one bit of information. So how many bits are there in the causally connected horizon region of our universe at this present epoch? The answer was worked out by Seth Lloyd (2002, 2006) using quantum mechanics. This is key: quantum mechanics says that the states of matter are fundamentally discrete rather than continuous, so they form a countable set. It is then possible to work out (approximately) how many bits of information any given volume of the universe contains by virtue of quantum discreteness. The answer is 10^{122} bits for the region within the horizon at this time. This number has a neat physical interpretation. It is the area of the horizon divided by the smallest area permitted by quantum discreteness, the so-called Planck area, $4\pi G\hbar/c^3$, which is roughly 10^{-65} cm². So the cosmic bit count is a dimensionless ratio, and a fundamental parameter of the universe.

Lloyd's number is not new in physical theory. It is roughly $N^{3/2}$, where N is the so-called Eddington–Dirac number: the ratio of electromagnetic to gravitational force between an electron and a proton. It is also the current age of the universe expressed in atomic units. Both Arthur Eddington (1931) and Paul Dirac (1937) attempted to build fundamental theories of physics using this number as the starting point. Neither attained any long-term success, so we must be careful to learn that lesson of history. However, Eddington and Dirac did not have the benefit of our better understanding of the relationship between gravitation and the concept of *entropy*. That understanding stemmed from important work done in the 1960s and 1970s on the physics of black holes. By 1970 it was obvious that black holes possess fundamental thermodynamic properties, and that the event horizon area of a black hole – roughly, the surface area of its boundary – plays the role of entropy. In standard thermodynamics, as applied to heat engines, say, entropy is a measure of the degree of disorder in a system, or, alternatively, the negative of the amount of useful energy that may be extracted to perform work. In the early 1970s, Jacob Bekenstein discovered that if quantum mechanics were applied to black holes, a specific expression could be given for its entropy (Bekenstein, 1973). This work was firmed up by Stephen Hawking (1975), who discovered that black holes are not perfectly black after all, but glow with heat radiation. The temperature of the radiation is inversely proportional to the mass M of the black hole, so that small black holes are hotter than large ones. The corresponding Bekenstein–Hawking entropy of an uncharged, non-rotating black hole is

$$S = 4\pi k G M^2 / \hbar c^3 = {}^1/_4 k A \qquad (4.1)$$

where A is its area in Planck units, and k is Boltzmann's constant, which converts units of energy to units of temperature. Significantly, in the black hole case entropy is a function of the boundary *area*, as opposed to volume. By contrast, the entropy of two masses of gas in identical thermodynamic states is the sum of the two *volumes* of gas.

I now come to the link with information. It has been known for many decades that entropy can be regarded as a measure of ignorance (Szilard, 1929, 1964). For example, if we know all the molecules of a mass of gas are confined to the corner of a box, the gas is ascribed a low entropy. Conversely, when the gas is distributed throughout the volume and its molecules are thoroughly shuffled and distributed chaotically, the entropy is high. Ignorance is the flip side of information, so we may deduce a mathematical relationship between entropy and information I. As given by Shannon (1948), that relationship is

$$S = -I \qquad (4.2)$$

One may think of the entropy of a gas as the information concerning the positions and motions of its molecules over which we have lost cognizance. In a similar vein, when matter falls into a black hole, we lose track of that too, because the black hole surface is an event horizon through which light cannot pass from the inside to the outside (which is why the hole is black). The Bekenstein–Hawking formula (4.1) relates the total information swallowed by a black hole to the surface area of its event horizon. The formula shows that the information of the black hole is simply one-quarter of the horizon area in Planck units.

The association of entropy and information with horizon area may be extended to *all* event horizons, not just those surrounding black holes; for example, the cosmological event horizon, which I discussed above (Davies and Davis, 2003; Gibbons and Hawking, 1977). Bekenstein proposed generalizing equation 4.1 to obtain a *universal* bound on entropy (or information content) for *any* physical system (Bekenstein, 1981). The black hole saturates the Bekenstein bound, and represents the maximum amount of information that can be packed into the volume encompassed by the horizon. A similar statement may be postulated for the cosmological horizon (where so-called de Sitter space saturates the bound).

The link between information (loss) and area seems to be a very deep property of the universe, and has been elevated to the status of a fundamental principle by

Gerhard 't Hooft (1993) and Leonard Susskind (1995), who proposed a so-called *holographic principle*, according to which the information content of a volume of space (any volume, not just a black hole) is captured by the information that resides on an enveloping surface that bounds that volume. (The use of the term "holographic" is an analogy based on the fact that a hologram is a three-dimensional image generated by shining a laser on a two-dimensional plate.) The holographic principle implies that the total information content of a region of space cannot exceed one-quarter of the surface area (in Planck units) that confines it (other variants of the holographic principle have been proposed, with different definitions of the enveloping area), and that this limit is attained in the case of the cosmological event horizon. If the holographic principle is applied to the state of the universe today, one recovers Lloyd's cosmic information bound of 10^{122} bits.

4.4 What does the finite information content of the universe tell us about "reality"?

The fact that (in the standard cosmological model at least) the information content of the universe is finite would seem to be a very important foundational fact about the universe. What are its implications? For a start, it means that nothing in the universe (as defined by the bounding horizon) can be specified or described by more than 10^{122} bits of information. By "nothing" I refer to actually-existing physical structures or states. The bound does not apply, for example, to merely hypothetical specifications, such as all possible hands of cards, or all possible combinations of amino acids making up a protein ($> 10^{130}$) because there is no claim that all such combinations might be physically present in the universe. Thus the universe could not contain a hotel with 10^{130} rooms, for example. In fact, the universe contains only about 10^{90} particles in total (including photons but not gravitons), and the finite information bound says they could not be confined to the "corner of a box" very much smaller than the universe, to borrow from the example of classical thermodynamics, because we would then know their locations to a better-than-permitted level of description. Note that the informational properties of the quantum universe differ fundamentally in this respect from the classical universe of Laplace's demon. Laplace assumed that the state of the universe at one instant could be specified to *infinite precision*; that is, the position and velocity of each particle could be ascribed a set of six real numbers. (It is easily shown that even tiny imprecisions lead to exponentially growing errors in the demon's prediction.) But almost all real numbers require an infinite amount of information to specify them.

The above-mentioned whimsical example (a hotel) refers to a classical state. How about a quantum state? After all, the information bound is quantum mechanical in nature. Consider a series of photon beam splitters labeled i, each of which permits a photon to traverse it (or be destroyed) with a certain probability p_i. Each encounter between a given photon and a beam splitter reduces the probability that the photon survives to exit the entire assemblage of beam splitters. After N such encounters, there is a probability $P(N) = p_1 p_2 p_3 \ldots p_N$ that the photon will have traversed the entire series. This becomes an exponentially small number as N rises. For example, if $p_i = \frac{1}{2}$ for all i and $N>400$, we find $2^{-N}<10^{-122}$. Can the universe contain such a small number? Of course it can in a sense: I just wrote it down! But how can we test if the prediction for the photon's penetration probability is correct? That is, how do we know quantum mechanics accurately describes this experimental set-up? We would have to perform $>10^{122}$ experiments to verify it, and that is certainly not only impossible for us, it is impossible *even in principle for a Laplace-type demon*. Now consider that the p_i are not all exactly $\frac{1}{2}$, but numbers chosen randomly from the interval $[0, 1]$. Then for almost all of the set $\{p_i\}$ the total probability $P(N)$ could *not* be accommodated in the universe. If a demon chose to write out the answer using every bit of information contained in the universe – every particle, say – the demon would run out of bits before the number could be expressed. Actually, the information would be very likely to be exhausted even for a *single* beam splitter, given that the real number p_1 could almost always be expressed only by stipulating an infinite number of digits: for example $0.37652583\ldots$

The question then arises, is the number $P(N)$ in some sense unknowable, not just in practice, but in principle? Expressed differently, could a demon even know that number? And if not – if the number is fundamentally unknowable – does that signal a fundamental limit in the level of precision at which quantum mechanics may be applied even in principle? The immediate answer to the latter question is no, because a *probability* is not an actuality; it is merely a relative weighing of actualities, and in that respect it possesses the same status as the number of possible combinations of amino acids. The demon (or for that matter a laboratory technician) can merely look to see whether or not the photon has survived, and the answer requires only one bit of information ("yes" or "no") to express it. But there is a subtlety buried here. Quantum mechanics generally cannot predict *actualities*, only *probabilities*. What it can predict – in principle with perfect accuracy – are wave-function amplitudes, from which the probabilities can be computed. In the example of the beam splitter, the wave function is a superposition of amplitudes, and the number of branches of the wave function, or components summed in the superposition, is 2^N. For $N>400$, this number alone exceeds the information-carrying capacity of the universe, never mind the information required to specify the amplitudes of each component of the superposition.

So my question now is, can the *quantum state* (the superposition of amplitudes that make up the wave function) be contained in the universe? The standard answer is yes. After all, what would prevent us from assembling 400 random beam splitters and sending in a photon? It is true that we can create such a state, but can we specify, or describe it? Presumably not – not even the demon can do that. Which brings me to the crux of the matter. Is the quantum state in any sense real, given that it is in principle unknowable from *within* the universe? Or is it merely a Platonic fiction, useful (as is the concept of infinity) for doing calculations, its fictional nature safely buried beneath very much larger experimental and initial condition errors? Given that a full specification or description of the beam-splitter experiment requires more information than the universe contains, my question becomes the following: is information something that *really exists*, independently of observers, or is it merely our description of what can in principle become known to an agent or observer? If the latter is the case – if information is merely a description of *what we know* about the physical world – then there is no reason why Mother Nature should care about the cosmic information bound, and no reason that the bound should affect fundamental physics. A Platonic Mother Nature can be all-knowing. And according to the orthodox view of laws, in which the bedrock of physical reality is vested in *perfect* laws of physics inhabiting the Platonic domain, Mother Nature can indeed compute to arbitrary precision with the unlimited quantity of information at her disposal. But if information is "real" – if, so to speak, it occupies the ontological basement (as I propose) – then the bound on the information content of the universe is a fundamental limitation on *all* of nature, and not just on states of the world that humans perceive.

A scientist who advocated precisely this position was Rolf Landauer, who adopted the view that "the universe computes in the universe," and not in a Platonic heaven, a point of view motivated by his insistence that "information is physical." Landauer was quick to spot the momentous consequences of this shift in perspective:

The calculative process, just like the measurement process, is subject to some limitations. A sensible theory of physics must respect these limitations, and should not invoke calculative routines that in fact cannot be carried out.   (Landauer, 1967)

In other words, in a universe limited in resources and time – for example, a universe subject to the cosmic information bound – concepts such as real numbers, infinitely precise parameter values, differentiable functions, and the unitary evolution of a wave function are a fiction: a useful fiction to be sure, but a fiction nevertheless. Consider the case of Laplace's demon, and the key phrase, "if this intellect were vast enough to submit the data to analysis." If Mother Nature – in effect, Laplace's demon – inhabits the Platonic realm of perfect, infinitely precise mathematics, then the finite information bound of the universe matters not at all, because the Platonic Mother Nature is, to paraphrase Laplace, certainly "vast enough," because she is omniscient and possesses *infinite* intellect, and can therefore submit an infinity of bits of data to analysis. She can indeed "carry out" the "calculative processes" to which Landauer refers. But if information is physical, if it is ontologically real and physically fundamental, then there *are* no Platonic demons, no godlike transcendent Mother Nature computing with real numbers; indeed, no real numbers. There is only the hardware of the real physical universe doing its own calculation itself, in the manner that Lloyd describes in Chapter 5 of this book. Expressed differently, the laws of physics are inherent in and emergent with the universe, not transcendent of it.

Landauer made his original comments as part of a general analysis; his ideas pre-dated the holographic principle and the finite information bound on the universe. But the existence of this bound projects Landauer's point beyond mere philosophy and places a real restriction on the nature of physical law. For example, one could not justify the application of the laws of physics in situations employing calculations that involve numbers greater than about 10^{122}, and if one did, then one might expect to encounter departures between theory and experiment. For most purposes, the information bound is such a large number that the consequences of the shift I am proposing are negligible. Consider the law of conservation of electric charge, for example. The law has been tested to about only one part in 10^{12}. If it were to fail at the 10^{122} bit level of accuracy, the implications are hardly significant.

Nevertheless there are situations in theoretical physics in which very large numbers do crop up. One obvious class of cases is where exponentiation occurs. Consider, for example, statistical mechanics, in which Poincaré's recurrence times are predicted to be of the order $\exp(10^{N})$ Planck times (chosen to make the number dimensionless) and N is the number of particles in the system. Imposing a bound of 10^{122} implies that the recurrence time prediction is reliable only for recurrence times of about 10^{60} years. Again, this is so long we would be unlikely to notice any departure between theory and observation.

A more striking and potentially practical application of the same principle is quantum computation. Quantum computers hold out the promise of possessing exponentially greater power than classical computers, because of the phenomena of quantum superposition and entanglement. The latter refers to the fact that two quantum systems, even when physically separated, are still linked in a subtle way. The arithmetic of the linkage reveals that there are exponentially more possible states of entangled quantum systems than their separate components contain. Thus an n-component system (for example, n atoms) possesses 2^n states, or 2^n components of the wave function describing the system. The fundamentally exponential character of the quantum realm has been eloquently addressed by Scott Aaronson (2005), using the pithy question: "Is the universe a polynomial or exponential place?"

to discuss what he calls "the ultimate Secret of Secrets."
He goes on to say:

For almost a century, quantum mechanics was like a Kabbalistic
secret that God revealed to Bohr, Bohr revealed to the physicists, and the physicists revealed (clearly) to no one. So long as
the lasers and transistors worked, the rest of us shrugged at all
the talk of complementarity and wave–particle duality, taking
for granted that we'd never understand, or need to understand,
what such things actually meant. But today – largely because of
quantum computing – the Schrödinger's cat is out of the bag,
and all of us are being forced to confront the exponential Beast
that lurks inside our current picture of the world. And as you'd
expect, not everyone is happy about that, just as the physicists
themselves weren't all happy when they first had to confront it
in the 1920s.   (Aaronson, 2005)

The question I now ask is whether Aaronson's "exponential Beast" is compatible with a Laplace-type demon
located *within* the real universe and subject to its finite
resources and age – a Laplacian demiurge would be a more
accurate description. Let me call this Beast "Landauer's
demon." Suppose it is required to predict the behavior of
a quantum computer subject to the above discussed cosmological information bound. The key to quantum computation lies with the exponential character of quantum
states, so here we have the crucial exponentiation at work
that is vulnerable to the cosmic information bound. To
be specific, a quantum state with more components than
about $n = 400$ particles is described by a wave function
with more components than Lloyd's 10^{122} bits of information contained in the entire universe. A generic wave
function of this state of 400-particles *could not be expressed*

in terms of bits of information, even in principle. Even if the entire universe were commandeered as a data display, it would not be big enough to accommodate the specification of that quantum state. So a generic 400-particle quantum state cannot be described, let alone its evolution predicted, even by a Landauer demon. It could, however, be predicted with a truly god-like transcendent Platonic demon with *infinite* resources and patience at its disposal.

The conclusion is stark. If the cosmic information bound is set at 10^{122} bits, and *if information is ontologically real*, then the laws of physics have intrinsically finite accuracy. For the most part, that limitation of the laws will have negligible consequences, but in cases of exponentiation, like quantum entanglement, they make a big difference, a difference that could potentially be observed. Creating a state of 400 entangled quantum particles is routinely touted by physicists working on building a quantum computer (their target is 10 000 entangled particles). I predict a breakdown of the unitary evolution of the wave function at that point, and possibly the emergence of new phenomena. To quote Wittgenstein (1921): "Whereof one cannot speak, thereof one must remain silent." We cannot – should not – pronounce on, or predict, the state, or dynamical evolution, of a generic quantum system with more than about 400 entangled particles, because there are not enough words in the entire universe to describe that state!

The orthodox position on the accuracy of predictions is that the laws of physics themselves are infinitely precise, but the concept of a perfectly isolated physical system and precisely known initial conditions are an idealization. In practice, so the argument goes, there will inevitably be errors, which will normally be enormously greater than one part in 10^{122}. In the case of quantum computation, these errors are tackled using error-correcting procedures and redundancy. The position I am advocating is that the finite information bound on the universe limits the accuracy of the laws themselves, rendering them irreducibly "fuzzy." This is a type of unavoidable cosmological noise, which no amount of error correction can remove. It would manifest itself as a breakdown in the unitary evolution of the wave function. What I am suggesting here seems to be close to the concept of unavoidable intrinsic decoherence proposed by Milburn (1991, 2006). Some clarification of these issues may emerge from the further study of the recent discovery that the entropy of quantum entanglement of a harmonic lattice also scales like area rather than volume (Cramer and Eisert, 2006), which would seem to offer support for the application of the holographic principle to entangled states. It would be good to know how general the entanglement–area relationship might be.

Finally, I should point out that the information bound (1) was derived using quantum field theory, but that same bound applies to quantum field theory. Ideally one should derive the bound using a self-consistent treatment. If one adopts the philosophy that information is primary and ontological, then such a self-consistency argument should be incorporated in a larger program directed at unifying mathematics and physics. If, following Landauer, one accepts that mathematics is meaningful only if it is the product of real computational processes (rather than existing independently in a Platonic realm) then there is a self-consistent loop: the laws of physics determine what can be computed, which in turn determines the informational basis of those same laws of physics. Paul Benioff (2002) has considered a scheme in which mathematics and the laws of physics co-emerge from a deeper principle of mutual self-consistency, thus addressing Wigner's famous question of why mathematics is so "unreasonably effective" in describing the physical world (Wigner, 1960). I have discussed these deeper matters elsewhere (Davies, 2006).

4.5 Conclusion

Let me finish with a personal anecdote. A vivid memory from my high-school years was learning how to calculate the gravitational potential energy of a point mass by integrating the (negative) work done transporting the particle radially inwards from infinity to the Earth's surface. I raised my hand and asked how a particle can actually be transported from infinity. The answer I received is that this mode of analysis is a device designed to make the calculation simple, and that the error involved in taking infinity as the starting point rather than some finite but very great distance is negligible. And so indeed it is. But this little exchange set me thinking about the use of mathematics as a device versus the actual mathematical nature of physical laws. I wanted to know whether the "real" potential energy of gravitation – the one based on the laws that Mother Nature herself uses – is the one that takes infinity as the starting point, or somewhere closer. In other words, although we human beings could never, even in principle, carry out the physical process involved in computing the exact potential energy, maybe nature somehow "knows" the answer without actually doing the experiment. That is, the exact answer is there, "embedded" in the laws of physics, and the business about infinity is a problem only for humans (and perhaps only for the likes of troublesome students such as the young Paul Davies).

The flip side of infinity is the use of infinitesimal intervals, which form the basis for the calculus. In view of the fact that all the fundamental laws of physics are expressed as differential equations, the status of infinitesimals is crucial. Again the question arises as to whether they are artifacts of human mathematics, or correspond to reality. A philosopher might express it by the question, What is the ontological status of infinitesimal intervals? This question is closely related to the use of real numbers and the property of continuity. Can an interval of space or time be subdivided without limit?

Over the years there have been many speculations by physicists that space–time may not in fact be continuous and differentiable, and that the application of real numbers and differentiation might be merely a convenient idealization. Most of these ideas remained restricted to small groups of devotees. The serious challenge came, however, not from physicists but computer scientists. Digital computers cannot, by their very nature, handle infinite and infinitesimal quantities because they manipulate discrete bits using Boolean algebra. This is in contrast to analog computers such as slide rules. The study of the performance of digital computers, which began with Turing, matured into an entire discipline in its own right in the decades after 1950. In recent years, theoretical physics and the theory of computation have been brought together, raising many deep conceptual issues that are as yet unresolved. The most basic of these is the following: Is the universe itself some sort of computer, and if so what? Analog, digital, quantum, or something else?

Lloyd advocates that the universe computes itself, but it does so quantum mechanically – the universe is a *quantum* computer (see Chapter 5 of this volume). We can envisage a corresponding quantum Landauer demon, able to observe all the branches of the wave function – in effect, all possible worlds, rather than a single actual world (this is precisely the ontology adopted in the Everett many-universes interpretation of quantum mechanics) (Everett, 1957). If that ontology is correct, if the ground of being is vested in the quantum wave function rather than the informational bits that emerge via measurement and observation, then the information bound on the universe is exponentially greater, and nothing would "go wrong" to an entangled state of 400 particles. Whether reality does lie in a quantum realm, to which human beings have no access, or whether it lies in the realm of real bits and real observations, could therefore be put to the test with a sufficiently complex quantum system. If the quantum computation optimists are right, it maybe that within a few years a new discipline will emerge: *experimental ontology*.

References

Aaronson, S. (2005). Are quantum states exponentially long vectors? *Proceedings of the Oberwolfach Meeting on Complexity Theory*, arXiv: quant-ph/0507242v1, accessed 8 March 2010 (http://arxiv.org/abs/quant-ph/0507242).

Barrow, J. D. (2002). *The Constants of Nature*. New York and London: Random House.

Bekenstein, J. D. (1973). Black holes and entropy. *Physical Review D*, 8: 2333–2346.

Bekenstein, J. D. (1981). Universal upper bound on the entropy-to-energy ratio for bounded systems. *Physical Review D*, 23: 287–298.

Benioff, P. (2002). Towards a coherent theory of physics and mathematics. *Foundations of Physics*, 32: 989–1029.

Boswell, J. (1823). *The Life of Samuel Johnson*, vol. 1. London: J. Richardson & Co.

Carroll, S. (2007). *Edge: The Third Culture*. Accessed 8 March 2010 (www.edge.org/discourse/science_faith.html).

Cramer, M., and Eisert, J. (2006). Correlations, spectral gap and entanglement in harmonic quantum systems on generic lattices. *New Journal of Physics*, 8: 71.

Davies, P. C. W., and Davis, T. M. (2003). How far can the generalized second law be generalized? *Foundations of Physics*, 32: 1877–1889.

Davies, P. (2006). *The Goldilocks Enigma: Why Is the Universe Just Right for Life?* London: Allen Lane, The Penguin Press.

Descartes, R. (1630). Letter to Mersenne, 15 April 1630. In *Descartes' Philosophical Letters*, trans. and ed. A. Kenny (1970). Oxford: Clarendon Press.

de Spinoza, B. (1670). *Theological–Political Treatise*, 2nd ed, trans. S. Shirley. Indianapolis, IN: Hackett Publishing, 75.

Dirac, P. A. M. (1937). The cosmological constants. *Nature*, 139: 323.

Drake, S. (1957). *Discoveries and Opinions of Galileo*. New York: Doubleday-Anchor.

Eddington, A. S. (1931). Preliminary note on the masses of the electron, the proton, and the universe. *Mathematical Proceedings of the Cambridge Philosophical Society*, 27: 15–19.

Everett, H. (1957). Relative state formulation of quantum mechanics. *Reviews of Modern Physics*, 29: 454–462.

Gibbons, G. W., and Hawking, S. W. (1977). Cosmological event horizons, thermodynamics, and particle creation. *Physical Review D*, 15: 2738–2751.

Greene, B. (1999). *The Elegant Universe*. New York: Norton.

Hawking, S. W. (1975). Particle creation by black holes. *Communications in Mathematical Physics*, 43: 199–220.

Hawking, S. (1988). *A Brief History of Time*. New York: Bantam.

Landauer, R. (1967). Wanted: A physically possible theory of physics. *IEEE Spectrum*, 4: 105–109, accessed 8 March 2010 (http://ieeexplore.ieee.org/xpl/freeabs_all.jsp?arnumber=5215588).

Landauer, R. (1986). Computation and physics: Wheeler's meaning circuit? *Foundations of Physics*, 16(6): 551–564.

Laplace, P. (1825). *Philosophical Essays on Probabilities*. Trans. F. L. Emory and F. W. Truscott (1985). New York: Dover.

Lloyd, S. (2002). Computational capacity of the universe. *Physical Review Letters*, 88: 237901.

Lloyd, S. (2006). *The Computational Universe*. New York: Random House.

Milburn, G. (1991). Intrinsic decoherence in quantum mechanics. *Physical Review A*, 44: 5401–5406.

Milburn, G. (2006). Quantum computation by communication. *New Journal of Physics*, 8: 30.

Rees, M. (2001). *Our Cosmic Habitat*. Princeton: Princeton University Press.

Russell, B. (1957). *Why I Am Not A Christian*. New York: Touchstone.

Shannon, C. E. (1948). A mathematical theory of communication. *Bell System Technical Journal*, 27: 379–423, 623–656.

Smolin, L. (2008). On the reality of time and the evolution of laws. Online video lecture and PDF, accessed 8 March 2010 (http://pirsa.org/08100049/).

Susskind, L. (1995). The world as a hologram. *Journal of Mathematical Physics*, 36: 6377.

Susskind, L. (2005). *The Cosmic Landscape: String Theory and the Illusion of Intelligent Design*. New York: Little, Brown.

Szilard, L. (1929). Über die Entropieverminderung in einem thermodynamischen System bei eingriffen intelligenter Wesen. *Zeitschrift für Physik*, 53: 840–856.

Szilard, L. (1964). On the decrease of entropy in a thermodynamic system by the intervention of intelligent beings. *Behavioral Science*, 9(4): 301–310.

Tegmark, M. (2003). Parallel universes. *Scientific American*, May 31, 2003.

't Hooft, G. (1993). Dimensional reduction in quantum gravity. arXiv: gr-qc/ 9310026v1, accessed 8 March 2010 (http:// arxiv.org/abs/gr-qc/9310026v1).

Ward, K. (1982). *Rational Theology and the Creativity of God*. New York: Pilgrim Press.

Wheeler, J. A. (1979). Frontiers of time. In *Problems in the Foundations of Physics*, ed. G. Toraldo di Francia. Amsterdam: North-Holland, 395–497.

Wheeler, J. A. (1983). On recognizing 'law without law'. *American Journal of Physics*, 51: 398–404.

Wheeler, J. A. (1989). Information, physics, quantum: The search for links. *Proceedings of the Third International Symposium on the Foundations of Quantum Mechanics* (Tokyo), 354.

Wheeler, J. A. (1994). *At Home in the Universe*. New York: AIP Press.

Wheeler, J. A., and Ford, K. (1998). It from bit. In *Geons, Black Holes & Quantum Foam: A Life in Physics*. New York: Norton.

Wigner, E. P. (1960). The unreasonable effectiveness of mathematics in the natural sciences. *Communications in Pure and Applied Mathematics*, 13(1): 1–14.

Wittgenstein, L. (1921). *Tractatus Logico-Philosophicus*, trans. D. Pears and B. McGuinness (1961). London: Routledge.

Zeilinger, A. (2004). Why the quantum? It from bit? A participatory universe? Three far-reaching, visionary questions from John Archibald Wheeler and how they inspired a quantum experimentalist. In *Science and Ultimate Reality: Quantum Theory, Cosmology, and Complexity*, eds Barrow J. D., Davies P. C. W., and C. L. Harper. Cambridge: Cambridge University Press, 201–220.

5

The computational universe

SETH LLOYD

~

It is no secret that we are in the midst of an information-processing revolution based on electronic computers and optical communication systems. This revolution has transformed work, education, and thought, and has affected the life of every person on Earth.

5.1 The information-processing revolutions

The effect of the digital revolution on humanity as a whole, however, pales when compared with the effect of the previous information-processing revolution: the invention of moveable type. The invention of the printing press was an information-processing revolution of the first magnitude. Moveable type allowed the information in each book, once accessible only to the few people who possessed the book's hand-copied text, to be accessible to thousands or millions of people. The resulting widespread literacy and dissemination of information completely transformed society. Access to the written word empowered individuals not only in their intellectual lives, but in their economic, legal, and religious lives as well.

Information and the Nature of Reality: From Physics to Metaphysics, eds. Paul Davies and Niels Henrik Gregersen. Published by Cambridge University Press
© P. Davies and N. Gregersen 2010, 2014.

Similarly, the effect of the printed word is small when compared with the effect of the written word. Writing – the discovery that spoken sounds could be put into correspondence with marks on clay, stone, or paper – was a huge information-processing revolution. The existence of complicated, hierarchical societies with extended division of labor depends crucially on writing. Tax records figure heavily in the earliest cuneiform tablets.

Just as printing is based on writing, writing stems from one of the greatest information-processing revolutions in the history of our planet: the development of the spoken word. Human language is a remarkable form of information processing, capable of expressing, well, anything that can be put into words. Human language includes within it the capacity to perform sophisticated analysis, such as mathematics and logic, as well as the personal calculations ("if she does this, I'll do that") that underlie the complexity of human society.

Although other animals have utterance, it is not clear that any of them possess the same capacity for universal language that humans do. Ironically, the entities that possess the closest approximation to human language are our own creations: digital computers, whose computer languages possess a form of universality bequeathed to them by human language. It is the social organization stemming from human language (together with written language, the printed word, computers, etc.) that have made human beings so successful as a species, to the extent that the majority of the planet's resources are now organized by humans for humans. If other species could speak, they would probably say, "Who ordered *that*?".

Before turning to even earlier information-processing revolutions, it is worth saying a few words about how human language came about. Who "discovered" human language? The fossil record, combined with recently revealed genetic evidence, suggests that human language may have arisen between 50 and 100 000 years ago, in Africa. Fossil skulls suggest that human brains underwent significant change over that time period, with the size of the cortex expanding tenfold. The result was our species, *Homo sapiens*: "man with knowledge" (literally, "man with taste"). Genetic evidence suggests that all women living today share mitochondrial DNA (passed from mother to daughter) with a single woman who lived in Africa around 70 000 years ago. Similarly, all men share a Y chromosome with one man who lived at roughly the same time.

What evolutionary advantage did this Adam and Eve possess over other hominids that allowed them to populate the world with their offspring? It is plausible that they possessed a single mutation or chance combination of DNA that allowed their offspring to think and reason in a new and more powerful way (Chomsky et al., 2002). Noam Chomsky has suggested that this way of reasoning should be identified with recursion, the ability to construct hierarchies of hierarchies, which lies at the root both of human language and of mathematical analysis. Once the ability to reason had appeared in the species, the theory goes, individuals who possessed this ability were better adapted to their immediate surroundings, and indeed, to all other surroundings on the planet. We are the offspring of those individuals.

Once you can reason, there is great pressure to develop a form of utterance that embodies that reason. Groups of *Homo sapiens* who could elaborate their way of speaking to reflect their reasoning would have had substantial evolutionary advantage over other groups who were incapable of complex communication and who were therefore unable to turn their thoughts into concerted action.

I present this plausible theory on the origin of language and of our species to show that information-processing revolutions need not be initiated by human beings. The "discovery" of a new way of processing information can arise organically out of an older way. Apparently, once the mammalian brain had evolved, then a few mutations gave rise to the ability to reason recursively. Once the powerful information-processing machinery of the brain was present, language could evolve by accident, coupled with natural selection.

Now let us return to the history of information-processing revolutions. One of the most revolutionary forms of information processing is *sex*. The original sexual revolution (not the one of the 1960s) occurred some billion years ago when organisms learned to share and exchange DNA. At first, sex might look like a bad idea: when you reproduce sexually, you never pass on your genome intact. Half of your DNA comes from your mother, and half from your father, all scrambled up in the process called recombination. By contrast, an asexually reproducing organism passes on its complete genome, modulo a few mutations. So even if you possess a truly fantastic combination of DNA, when you reproduce sexually, your offspring may not possess that combination. Sex messes with success.

So why is sex a good idea? Exactly because it scrambles up parents' DNA, sexual reproduction dramatically increases the potential rate of evolution. Because of the scrambling involved in recombination, sexual reproduction opens up a huge variety of genetic combinations for your offspring, combinations that are not available to organisms that rely on mutation alone to generate genetic variation. (In addition, whereas most mutations are harmful, recombination assures that viable genes are recombined with other viable genes.) To compare the two forms of reproduction, sexual and asexual, consider the following example: by calculating the number of genetic combinations that can be generated, it is not hard to show that a small town of 1000 people, reproducing sexually with a generation time of 30 years, produces the same amount of genetic variation as a culture of one trillion bacteria, reproducing asexually every 30 minutes.

Sex brings us back to the mother of all information-processing revolutions: life itself. However it came about, the mechanism of storing genetic information in DNA, and reproducing with variation, is a truly remarkable "invention" that gave rise to the beautiful and rich world around us. What could be more majestic and wonderful? Surely, life is the original information-processing revolution.

Or is it? Life arose on Earth some time in the last five billion years (for the simple reason that the Earth itself has only been around for that long). Meanwhile, the universe itself is a little less than fourteen billion years old. Were the intervening nine billion years completely devoid of information-processing revolutions?

The answer to this question is "No." Life is not the original information-processing revolution. The very first information-processing revolution, from which all other revolutions stem, began with the beginning of the universe itself. The big bang at the beginning of time consisted of huge numbers of elementary particles, colliding at temperatures of billions of degrees. Each of these particles carried with it bits of information, and every time two particles bounced off each other, those bits were transformed and processed. The big bang was a bit bang. Starting from its very earliest moments, every piece of the universe was processing information. The universe computes. It is this ongoing computation of the universe itself that gave rise naturally to subsequent information-processing revolutions such as life, sex, brains, language, and electronic computers.

5.2 The computational universe

The idea that the universe is a computer might at first seem to be only a metaphor. We build computers. Computers are the defining machines of our era. Consequently, we declare the universe to be a computer, in the same way that the thinkers of the Enlightenment declared the universe to be a clockwork one. There are two responses to this assertion that the computational universe is a metaphor. The first response is that, even taken as a metaphor, the mechanistic paradigm for the universe has proved to be incredibly successful. From its origins almost half a millennium ago, the mechanistic paradigm has given rise to physics, chemistry, and biology. All of contemporary science and engineering comes out of the mechanistic paradigm. To think of the universe not just as a machine, but also as a machine that computes, is a potentially powerful extension of the mechanistic paradigm.

The second response is that the claim that the universe computes is literally true. In fact, the scientific demonstration that all atoms and elementary particles register bits of information, and that every time two particles collide those bits are transformed and processed, was given at the end of the nineteenth century, long before computers occupied people's minds. Beginning in the 1850s, the great statistical mechanicians James Clerk Maxwell in Cambridge and Edinburgh, Ludwig Boltzmann in Vienna, and Josiah Willard Gibbs at Yale, derived the mathematical formulae that characterized the physical quantity known as entropy (Ehrenfest and Ehrenfest, 2002). Prior to their work, entropy was known as a somewhat mysterious thermodynamic quantity that gummed up the works of steam engines, preventing them from doing as much work as they otherwise might do. Maxwell, Boltzmann, and Gibbs wanted to find a definition of entropy in terms of the microscopic motions of atoms. The formulae that they derived showed that entropy was proportional to the number of bits of information registered by those atoms in their motions. Boltzmann then derived his eponymous equation to describe how those bits were transformed and flipped when atoms collide. At bottom, the universe is processing information.

The scientific discovery that the universe computes long preceded the formal and practical idea of a digital computer. It was not until the mid twentieth century, however, with the work of Claude Shannon and others, that the interpretation of entropy as information became clear (Shannon and Weaver, 1963). More recently, in the 1990s, researchers showed just how atoms and elementary particles compute at the most fundamental level (Chuang and

Nielsen, 2000). In particular, these researchers showed how elementary particles could be programmed to perform conventional digital computations (and, as will be discussed below, to perform highly unconventional computations as well). That is, not only does the universe register and process information at its most fundamental level, as was discovered in the nineteenth century, it is literally a computer: a system that can be programmed to perform arbitrary digital computations.

You may ask, So what? After all, the known laws of physics describe the results of experiments to exquisite accuracy. What does the fact that the universe computes buy us that we did not already know?

The laws of physics are elegant and accurate, and we should not discard them. Nonetheless, they are limited in what they explain. In particular, when you look out your window you see plants and animals and people; buildings, cars, and banks. Turning your telescope to the sky you see planets and stars, galaxies and clusters of galaxies. Everywhere you look, you see immense variation and complexity. Why? How did the universe get this way? We know from astronomical observation that the initial state of the universe, fourteen billion years ago, was extremely flat, regular, and simple. Similarly, the laws of physics are simple: the known laws of physics could fit on the back of a T-shirt. Simple laws, simple initial state. So where did all of this complexity come from? The laws of physics are silent on this subject.

By contrast, the computational theory of the universe has a simple and direct explanation for how and why the universe became complex. The history of the universe in terms of information-processing revolutions, each arising naturally from the previous one, already hints at why a computing universe necessarily gives rise to complexity. In fact, we can prove mathematically that a universe that computes must, with high probability, give rise to a stream of ever-more-complex structures.

5.3 Quantum computation

In order to understand how and why complexity arises in a computing universe, we must understand more about how the universe processes information at its most fundamental scales. The way in which the universe computes is governed by the laws of physics. Quantum mechanics is the branch of physical law that tells us how atoms and elementary particles behave, and how they process information.

The most important thing to remember about quantum mechanics is that it is strange and counterintuitive. Quantum mechanics is weird. Particles correspond to waves; waves are made up of particles; electrons and basketballs can be in two places at once; elementary particles exhibit what Einstein called "spooky action at a distance." Niels Bohr, one of the founders of quantum mechanics, once said that anyone who can contemplate quantum mechanics without getting dizzy has not properly understood it.

This intrinsically counterintuitive nature of quantum mechanics explains why many brilliant scientists, notably Einstein (who received his Nobel prize for his work in quantum mechanics), have distrusted the field. More than others, Einstein had the right to trust his intuition. Quantum mechanics contradicted his intuition, just as it contradicts everyone's intuition. So Einstein thought quantum mechanics could not be right: "God doesn't play dice," he declared. Einstein was wrong. God, or whoever it is who is doing the playing, plays dice.

It is this intrinsically chancy nature of quantum mechanics that is the key to understanding the computing universe. The laws of physics clearly support computation: I am writing these words on a computer. Moreover, physical law supports computation at the most fundamental levels: Maxwell, Boltzmann, and Gibbs show that all atoms register and process information. My colleagues and I exploit this information-processing ability of the universe to build quantum computers that store and process information at the level of individual atoms. But who – or what – is programming this computing computer? Where do the bits of information come from that tell the universe what to do? What is the source of all the variation and complexity that you see when you look out your window? The answer lies in the throws of the quantum dice.

Let us look more closely at how quantum mechanics injects information into the universe. The laws of quantum mechanics are largely deterministic: most of the time, each state gives rise to one, and only one, state at a later time. It is this deterministic feature of quantum mechanics that allows the universe to behave like an ordinary digital computer, which processes information in a deterministic fashion. Every now and then, however, an element of chance is injected into quantum evolution: when this happens, a state can give rise probabilistically to several different possible states at a later time. The ability to give rise to several different possible states allows the universe to behave like a quantum computer, which, unlike a conventional digital computer, can follow several different computations simultaneously.

The mechanism by which quantum mechanics injects an element of chance into the operation of the universe is called "decoherence" (Gell-Mann and Hartle, 1994). Decoherence effectively creates new bits of information, bits which previously did not exist. In other words, quantum mechanics, via decoherence, is constantly injecting new bits of information into the world. Every detail that we see around us, every vein on a leaf, every whorl on a fingerprint, every star in the sky, can be traced back to some bit that quantum mechanics created. Quantum bits program the universe.

Now, however, there seems to be a problem. The laws of quantum mechanics imply that the new bits that decoherence injects into the universe are essentially random, like the tosses of a fair coin: God plays dice. Surely, the universe did not arise completely at random! The patterns that we see when we look out the window are far from random. On the contrary, although detailed and complex, the information that we see around us is highly ordered. How can highly random bits give rise to a detailed, complex, but orderly universe?

5.4 Typing monkeys

The computational ability of the universe supplies the answer to how random bits necessarily give rise to order and complexity. To understand how the combination of randomness together with computation automatically gives rise to complexity, first look at an old and incorrect explanation of the origin of order and complexity. Could the universe have originated from randomness alone? No! Randomness, taken on its own, gives rise only to gibberish, not to structure. Random information, such as that created by the repeated flipping of a coin, is highly unlikely to exhibit order and complexity.

The failure of randomness to exhibit order is embodied in the well-known image of monkeys typing on typewriters, created by the French mathematician Emile Borel in the first decade of the twentieth century (Borel, 1909). Imagine a million typing monkeys (*singes dactylographiques*), each typing characters at random on a typewriter. Borel noted that these monkeys had a finite probability of producing all the texts in all the richest libraries of the world. He then pointed out that the chance of them doing so was infinitesimally small. (This image has appeared again and again in popular literature, as in the story that the monkeys immediately begin to type out Shakespeare's *Hamlet*.)

To see how small a chance the monkeys have of producing any text of interest, imagine that every elementary particle in the universe is a "monkey," and that each particle has been flipping bits or "typing," since the beginning of the universe. Elsewhere, I have shown that the number of elementary events or bit flips that have occurred since the beginning of the universe is no greater than $10^{120} \approx 2^{400}$. If one searches within this huge, random bit string for a specific substring (for example, Hamlet's soliloquy), one can show that the longest bit string that one can reasonably expect to find is no longer than the logarithm of the length of the long, random string. In the case of the universe, the longest piece of Hamlet's soliloquy one can expect to find is 400 bits long. To encode a typewriter character such as a letter takes seven bits. In other words, if we ask the longest fraction of Hamlet's soliloquy that monkeys could have produced since the beginning of the universe, it is, "To be, or not to be – that is the question: Whether 'tis nobler..." Monkeys, typing at random into typewriters, would not produce Hamlet, let alone the complex world we see around us.

Now suppose that, instead of typing on typewriters, the monkeys type their random strings of bits into computers. The computers interpret each string as a program, a set of instructions to perform a particular computation. What then? At first it might seem that random programs should give rise to random outputs: garbage in, garbage out, as computer scientists say. At second glance, however, one finds that there are short, seemingly random programs that instruct the computer to do all kinds of interesting things. (The probability that monkeys typing into a computer produce a given output is the subject of the branch of mathematics called algorithmic information theory.) For example, there is a short program that instructs the computer to calculate the digits of π, and a second program that instructs the computer to construct intricate fractal patterns. One of the shortest programs instructs the computer to compute all possible mathematical theorems and patterns, including every pattern ever generated by the laws of physics! One might say that the difference between monkeys typing into typewriters and monkeys typing into computers is all the difference in the world.

To apply this purely mathematical construct of algorithmic information theory to our universe, we need two ingredients: a computer, and monkeys. But we have a computer – the universe itself, which at its most microscopic level is busily processing information. Where are the monkeys? As noted above, quantum mechanics provides the universe with a constant supply of fresh, random bits, generated by the process of decoherence. Quantum fluctuations are the "monkeys" that program the universe (Lloyd, 2006).

To recapitulate:

(1) The mathematical theory of algorithmic information implies that a computer that is supplied with a random program has a good chance of producing all the order and complexity that we see. This is simply a mathematical fact: to apply it to our universe we need to identify the computing mechanism of the universe, together with its source of randomness.

(2) It has been known since the end of the nineteenth century that if the universe can be regarded as a machine (the mechanistic paradigm), it is a machine that processes information. In the 1990s, I and other researchers in quantum computation showed that the universe was capable of full-blown digital computation at its most microscopic levels: the universe is, technically, a giant quantum computer.

(3) Quantum mechanics possesses intrinsic sources of randomness (God plays dice) that program this computer. As noted in the discussion of the history of information-processing revolutions above, the injection of a few random bits, as in the case of genetic mutation or recombination, can give rise to a radically new paradigm of information processing.

5.5 Discussion

The computational paradigm for the universe supplements the ordinary mechanistic paradigm: the universe is not just a machine, it is a machine that processes information. The universe computes. The computing universe is not a metaphor, but a mathematical fact: the universe is a physical system that can be programmed at its most microscopic level to perform universal digital computation. Moreover, the universe is not just a computer: it is a quantum computer. Quantum mechanics is constantly injecting fresh, random bits into the universe. Because of its computational nature, the universe processes and *interprets* those bits, naturally giving rise to all sorts of complex order and structure (Lloyd, 2006).

The results of the previous paragraphs are scientific results: they stem from the mathematics and physics of information processing. Aristotle, when he had finished writing his *Physics*, wrote his *Metaphysics*: literally "the book after physics." This chapter has discussed briefly the physics of the computing universe and its implications for the origins of complexity and order. Let us use the physics of the computing universe as a basis for its metaphysics.

References

Borel, E. (1909). *Éléments de la Théorie des Probalités*. Paris: A. Hermann et Fils.

Chomsky, N., Hauser, M. D., and Tecumseh Fitch, W. (2002). The faculty of language: What is it, who has it, and how did it evolve. *Science*, 22(2): 1569–1579.

Chuang, I. A., and Nielsen, M. A. (2000). *Quantum Computation and Quantum Information*. Cambridge, UK: Cambridge University Press.

Ehrenfest, P., and Ehrenfest, T. (2002). *The Conceptual Foundations of the Statistical Approach in Mechanics*. New York: Dover.

Gell-Mann, M., and Hartle, J. B. (1994). *The Physical Origins of Time Asymmetry*, ed. J. Halliwell, J. Pérez-Mercader, and W. Zurek. Cambridge, UK: Cambridge University Press.

Lloyd, S. (2006). *Programming the Universe*. New York: Knopf.

Shannon, C. E., and Weaver, W. (1963). *The Mathematical Theory of Communication*. Urbana: University of Illinois Press.

6

Minds and values in the quantum universe

HENRY STAPP

~

Copenhagen is the perfect setting for our discussion of matter and information. We have been charged 'to explore the current concept of matter from scientific, philosophical, and theological perspectives'. The essential foundation for this work is the output of the intense intellectual struggles that took place here in Copenhagen during the twenties, principally between Niels Bohr, Werner Heisenberg, and Wolfgang Pauli. Those struggles replaced the then-prevailing Newtonian idea of matter as 'solid, massy, hard, impenetrable, moveable particles' with a new concept that allowed, and in fact demanded, entry into the laws governing the motion of matter of the consequences of decisions made by human subjects. This change in the laws swept away the meaningless billiard-ball universe, and replaced it with a universe in which we human beings, by means of our intentional effort, can make a difference in how the 'matter' in our bodies behaves.

Information and the Nature of Reality: From Physics to Metaphysics, eds. Paul Davies and Niels Henrik Gregersen. Published by Cambridge University Press © P. Davies and N. Gregersen 2010, 2014.

6.1 The role of mind in nature

Unfortunately, most of the prevailing descriptions of quantum theory tend to emphasize puzzles and paradoxes in a way that makes philosophers, theologians, and even non-physicist scientists leery of actually using in any deep way the profound changes in our understanding of human beings in nature wrought by the quantum revolution. Yet, properly presented, quantum mechanics is thoroughly in line with our deep human intuitions. It is the 300 years of indoctrination with basically false ideas about how nature works that now makes puzzling a process that is completely in line with normal human intuition. I therefore begin with a non-paradox-laden description of the quantum universe and the place of our minds within it.

The founders of quantum mechanics presented this theory to their colleagues as essentially a set of rules about how to make predictions pertaining to what we human observers will see, or otherwise experience, under certain specified kinds of conditions. Classical mechanics can, of course, be viewed in exactly the same way, but the two theories differ profoundly in the nature of the predictions they make.

In classical mechanics, the state of any system – at some fixed time, t – is defined by giving the location and the velocity of every particle in that system, and by giving also the analogous information about the fields. All observers and their acts of observation are simply parts of the evolving, fully predetermined, physically described universe. Within that framework the most complete prediction pertaining to any specified time is simply the complete description of the state of the universe at that time.

This complete description is in principle predictable in terms of the laws of motion and the complete description of the state of the universe at any other time.

Viewed from this classical perspective, even the *form* of the predictions of quantum mechanics seems absurd. The basic prediction of quantum theory is an answer to a question of the following kind: If the state of some system immediately before time *t* is the completely specified state #1, then what is the probability of obtaining the answer 'Yes' if we perform at time *t* an experiment that will reveal to us whether or not the system is in state #2?

Classical physics gives a simple answer to this question: the predicted probability is unity or zero, according to whether or not state #2 is the same as state #1. But quantum theory gives an answer that generally is neither unity nor zero, but some number in between.

The quantum structure is easily understood if we follow Heisenberg's idea of introducing the Aristotelian idea of *potentia*. A *potentia*, in Heisenberg's words, is an 'objective tendency' for some event to happen. Everything falls neatly in place if we assert – or simply recognize – that the quantum state of a system specifies the 'objective tendency' for a quantum event to happen, where a *quantum event* is the occurrence of some particular outcome of some particular action performed upon the system. In short, the quantum state is best conceived *in principle* exactly as it is conceived *in actual practice*, namely as a compendium of the objective tendencies for the appearances of the various physically possible outcomes of the various physically possible probing actions. Once the action to be performed upon the system is selected, the objective tendencies are expressed as probabilities assigned to the various alternative possible outcomes of that chosen action.

If one accepts as fundamental this Aristotelian idea of *potentia* – of objective tendencies – then the whole scheme of things becomes intuitively understandable. There is nothing intrinsically incomprehensible about the idea of 'tendencies'. Indeed, we build our lives upon this concept. However, three centuries of false thinking has brought many physicists and philosophers to expect and desire an understanding of nature in which everything is completely predetermined in terms of the physically described aspects of nature *alone*. Contemporary physics violates that classical ideal. Bowing partially to advances in physics, these thinkers have accepted the entry of 'randomness' – of mathematically controlled chance – as something they can abide, and even embrace. However, there remains something deeply galling to minds attuned to the conception of nature that reigned during the eighteenth and nineteenth centuries. This is the possibility that our human minds can introduce elements of definiteness into the description of nature that the physically described processes of nature, acting alone, leave unspecified.

The causal incompleteness of the physically described aspects of nature entailed by this possibility is something that many physicists, philosophers, and neuroscientists will go to extreme lengths to try to circumvent. Yet, such attempts all boil down, at present, to 'promissory materialism', for no one has yet shown how the interventions of our minds – or some surrogates – required by contemporary orthodox quantum theory can consistently be eliminated.

This seemingly unavoidable entry of mental realities into the laws of physics arises in connection with the choice of which (probing) action will be performed on a system being observed. Quantum theory places no conditions, statistical or otherwise, on these choices. Consequently, there is a 'causal gap' in orthodox contemporary physics. This gap is not in the choice of an *outcome*, which is mathematically controlled, at least statistically. It is rather in the choice of which of the physically possible probing actions will be undertaken. But the choices of which actions a person will make are exactly the choices that are important to religion, and more generally to moral philosophy. Thus orthodox contemporary physical theory offers a conception of nature that enforces, in a rationally coherent and massively confirmed way, everything that physics says about the structure of human experience, while leaving open the vitally important question of how we choose our actions from among the possibilities proffered by the causally incomplete physical laws.

6.2 The large and the small

A main source of confusion in the popular conception of quantum mechanics is a profound misunderstanding of the connection between the large and the small. One repeatedly hears the mantra 'quantum mechanics concerns very small things, whereas consciousness is related to large-scale activities in the brain; therefore quantum mechanics is not relevant to the problem of the connection between mind and brain'.

Actually, the basic problem resolved by orthodox quantum theory is precisely the problem of the connection between invisible small-scale activities and the large-scale activities that are more directly connected to conscious experience. The atomic-scale aspects of nature considered alone are unproblematic: they are governed by local deterministic laws that are well understood and logically manageable. It is only when the consequences of the atomic-level processes are extended to the macro level (for example, to Schrödinger's cat, or to Geiger counters, or to human brains) that the radically new quantum features come into play. It is only *then* that one encounters the seismic shift from a continuous deterministic process to the Heisenberg/Aristotelian notion of the *potentia* for psychophysical events to occur.

The deeper aspects of quantum theory concern precisely the fact that the purely physical laws of motion that work so well on the atomic scale fail to account for the *observed properties* of large conglomerations of atoms. It is *exactly* this problem of the connection between physically described small-scale properties and directly experienced large-scale properties that orthodox quantum theory successfully resolves. To ignore this solution, and cling to the false precepts of classical mechanics that leave mind and consciousness completely out of the causal loop, seems to be totally irrational. What fascination with the weird and the incredible impels philosophers to adhere, on the one hand, to a known-to-be-false physical theory that implies that *all* of our experiences of our thoughts influencing our actions are *illusions*, and to reject, on the other hand, the offerings of its successor, which naturally produces an image of ourselves that is fully concordant with our normal intuitions, and can explain how bodily behaviour can be influenced by *felt evaluations* that emerge from an aspect of reality that is not adequately conceptualized in terms of the mechanistic notion of bouncing billiard balls?

6.3 Decoherence

Decoherence effects are often cited as another reason why quantum effects cannot be relevant to an understanding of the mind–brain connection. Actually, however, decoherence effects are the basis both of the mechanism whereby our thoughts can affect our actions, and of the reconciliation of quantum theory with our basic intuitions.

The interaction of the various parts of the brain with their environment has the effect of reducing an extremely complex conceptualization of the state of the brain to something everybody can readily understand. The quantum state of the brain is reduced by these interactions to a collection of *parallel potentialities*, each of which is essentially a classically conceivable possible state of the brain. The word 'essentially' highlights the fact that each of the classical possibilities must be slightly smeared out to bring it into accord with Heisenberg's uncertainty principle: the potential location and velocity of the centre of each particle is smeared out over a small region. This conception of the quantum brain is intuitively accessible, and it is made possible by (environment-induced) decoherence. This picture of the brain captures very well the essence of the underlying mathematical structure, and it can be used with confidence.

According to this picture, your physically described brain is an evolving cloud of essentially classically conceivable potentialities. Owing to the uncertainty principle smearing, this cloud of potentialities can quickly expand to include the neural correlates of many mutually exclusive possible experiences. Each human experience is an aspect of a psycho-physical event whose psychologically described aspect is that experience itself, and whose physically described aspect is the reduction of the cloud of potentialities to those that contain the neural correlate of that experience.

These psycho-physical actions/events are of two kinds. An action of the first kind is a choice of how the observed system is to be probed. Each such action decomposes the continuous cloud of potentialities into a set of mutually exclusive but collectively exhaustive separate components. An action of the second kind is a choice 'on the part of nature' of which of these alternative possible potentialities will be 'actualized'. The actions of the second kind are predicted to conform to certain quantum probability rules. An action of the first kind is called by Bohr 'a free choice on the part of the experimenter'. It is controlled by no known law or rule, statistical or otherwise.

Decoherence plays also another crucial role. The rules of quantum theory allow a person to pose the question, 'Is my current state the same as a certain one that I previously experienced?' In the important cases in which a person wants to perform a certain action, the freedom of choice allowed by quantum theory permits that person to ask the question, 'Is my current state the one that on previous occasions usually led to a feedback that indicated the successful performance of that action?' The answer may or may not be 'Yes'. If the answer is 'Yes', then the neural correlate of 'Yes' will be actualized. It is reasonable to assume that this neural correlate will be a large-scale pattern of brain activity that, if actualized, will tend to cause the intended action to occur.

This tendency can be reinforced by exploiting the person's capacity – within the framework of the quantum laws – *to pose a question of his or her own choosing at any time of his or her own choosing.* This freedom can be used to activate a decoherence effect called the quantum Zeno effect. This effect can cause the neural correlate of the 'Yes' outcome to be held in place for longer than would otherwise be the case, provided the intentional effort to perform the action causes the same question to be posed repeatedly in sufficiently rapid succession. The freedom to do this is allowed by the quantum rules. The quantum Zeno effect is a decoherence effect, and it is not diminished by the environment-induced decoherence: it survives intact in a large, warm, wet brain.

The upshot of all this is that the arguments that were supposed to show why quantum mechanics is not relevant to the mind–body problem all backfire, and end up supporting the viability of a quantum mechanical solution that is completely in line with our normal intuitions. One need only accept what orthodox quantum mechanics insists upon – to the extent that it goes beyond an agnostic or pragmatic stance – namely that the physically described world is not a world of material substances, as normally conceived, but is rather a world of potentialities for future experiences.

Given this new playing field, we may commence dialogues pertaining to the remaining, and vitally key, issue: namely the origin and significance of the felt evaluations that seem to guide our actions. These evaluations appear to come from an experiential or spiritual realm, and are certainly allowed by quantum theory to have the effects that they seem to have. But before turning to these core questions, it may be useful first to elaborate upon some aspects of the preceding remarks.

6.4 The intuitive character of quantum theory

I claimed above that quantum mechanics, properly presented, and more specifically the quantum mechanical conception of nature, is in line with intuition. It is rather classical physics that is non-intuitive. It is only the viewing the quantum understanding of nature from the classical perspective, generated by three centuries of indoctrination, that makes the quantum conception appear non-intuitive.

Some other speakers, following common opinion, have said just the opposite.

Ernan McMullin has given, in Chapter 2 of this volume, a brief account of the history, in philosophy and in physics, of the meaning of 'matter'. Aristotle introduced essentially this term in connection with the notion of 'materials for making', such as timber. The Neo-Platonists used it in contrast to the 'spiritual' aspects of reality. In the seventeenth, eighteenth, and nineteenth centuries it became used to denote the carrier of the small set of properties that, according to the then-ascendant 'mechanical philosophy', were the only properties that were needed to account for all changes in the visible world. These properties, called 'physical properties', were considered to be 'objective', in contrast to the 'subjective properties', which are 'dependent in one way or another on the perceiver'.

In Chapter 2 of this volume, McMullin describes the two millennia of philosophical wanderings and wonderings about the 'stuff' out of which nature was built that occurred between the time of the Ionian philosophers and the invention of the classical conception by Isaac Newton. This account makes clear the fact that the classical conception of nature is not the direct product of innate human intuition. Schoolchildren need to be *taught* that the solid-looking table is 'really' mostly empty space, in which tiny atomic particles are buzzing around. And this conception leaves unanswered – and unanswerable in any way that builds rationally upon that classical conception – the question: How do our subjective experiences of the visible properties emerge from this conceptually and causally self-sufficient classically conceived reality?

The deepest human intuition is not the immediate grasping of the classical-physics-type character of the external world. It is rather that one's own conscious subjective efforts can influence the experiences that follow. Any conception of nature that makes this deep intuition an illusion is counterintuitive. Any conception of reality that cannot explain how our conscious efforts influence our bodily actions is problematic. What is actually deeply intuitive is the continually reconfirmed fact that our conscious efforts can influence certain kinds of experiential feedback. A putatively rational scientific theory needs at the very least to explain this connection in a rational way to be in line with intuition.

As regards the quantum mechanical conception, McMullin calls it 'problematic' and 'counterintuitive'. In Chapter 5 of this volume, Seth Lloyd calls it 'counterintuitive' and 'weird'. Let me explain why the opposite is true: why contemporary opinion, to the contrary, is the product of a distorted viewpoint that is itself counterintuitive, but has, in spite of its serious technical failings and inadequacies, been pounded into 'informed' human thinking by 300 years of intense indoctrination.

The original (Copenhagen) interpretation of quantum theory was pragmatic and epistemological: it eschewed ontology. It avoided all commitments about what really exists! Von Neumann retained and rigorized the essential mathematical precepts of the Copenhagen interpretation but, by developing the mind–matter parallelism of the Copenhagen conception, brought the bodies and brains of the human observer/experimenter into the world understood to be made of atoms, molecules, and the like. Von Neumann's formulation (called 'the orthodox interpretation' by Wigner) prepared the way for an imbedding ontology. This extension made by von Neumann is the basis of all attempts by physicists to go beyond the epistemological/pragmatic Copenhagen stance, and give an account of the reality that lies behind the phenomena.

Bohr sought to provide an adequate understanding of quantum theory, and our place within that understanding, that stayed strictly within the epistemological framework. Heisenberg, however, was willing to opine about 'what was really happening'.

Reality, according to Heisenberg, is built not out of matter, as matter was conceived of in classical physics, but out of psycho-physical *events* – events with certain aspects that are described in the language of psychology and with other aspects that are described in the mathematical language of physics – and out of *objective tendencies* for such events to occur. 'The probability function... represents a tendency for events and our knowledge of events' (Heisenberg, 1958, p. 46). 'The observation... enforces the description in space and time but breaks the determined continuity by changing our knowledge' (pp. 49–50). 'The transition from the "possible" to the "actual" takes place during the act of observation. If we want to describe what happens... we have to realize that the word "happens" can apply only to the observation, not to the state of affairs between two observations' (p. 54). 'The probability function combines objective and subjective elements. It contains statements about possibilities or better tendencies (*potentia* in Aristotelian philosophy), and these statements are completely objective: they do not depend on any observer; and it contains statements about our knowledge of the system, which of course are subjective, in so far as they may be different for different observers' (p. 53).

Perhaps the most important change in the theory, vis-à-vis classical physics, was its injection of the thoughts and intentions of the human experimenter/observer into the physical dynamics: 'As Bohr put it... in the drama of existence we ourselves are both players and spectators... our own activity becomes very important' (Heisenberg, 1958, p. 58). 'The probability function can be connected to reality only if one essential condition is fulfilled: if a new measurement is made to determine a *certain* property of the system' (p. 48 [my italics]). Bohr: 'The freedom of experimentation... corresponds to the free choice of experimental arrangement for which the mathematical structure of the quantum mechanical formalism offers the appropriate latitude' (Bohr, 1958, p. 73).

This 'choice on the part of the "observer"' is represented in the mathematical formalism by von Neumann's 'process 1' intervention (von Neumann, 1955, pp. 351, 418). It is the first – and absolutely essential – part of the process leading up to the final actualization of a new 'reduced' state of the system being probed by the human agent. This process 1 action partitions the existing state, which represents a continuous smear of potentially experienceable possibilities into a (countable) set of experientially distinct possibilities. There is nothing known in the mathematical description that determines the specifics of this *logically needed* reduction of a continuum to a collection of distinct possibilities, each associated with a different possible increment of knowledge. Also, the *moment at which* a particular process 1 action occurs is not specified by the orthodox quantum mathematical formalism. This choice of *timing* is part of what seems to be, and in actual practice is, determined by the observer's free choice. These basic features of quantum mechanics provide the basis for a rational and natural quantum dynamical explanation of how a person's conscious effortful intents can affect his or her bodily actions (Beauregard, Schwartz, and Stapp, 2005; Stapp, 2005; Stapp, 2006).

Many of our conscious experiences are associated with a certain element of intent and effort, and it may be that *every* conscious experience, no matter how spontaneous or passive it seems to be, has *some* degree of focusing of attention associated with it. An increase in the effortful intention associated with a thought intensifies the associated experience. Hence it is reasonable to assume that an application of effort increases the repetition rate of a sequence of essentially equivalent events.

If the rapidity of the process 1 events associated with a given intent is great enough, then, as a direct consequence of the quantum laws of change, the neural correlate of that intent will become almost frozen in place. This well-known and much studied effect is called the 'quantum Zeno effect'.

The neural correlate of an intent to act in a certain way would naturally be a pattern of neural activity that tends to cause the intended action to occur. Holding this pattern in place for an extended period ought strongly to tend to make that action occur. Thus a prominent and deeply appreciated gap in the dynamical completeness of orthodox quantum mechanics can be filled in a very natural way that renders our conscious efforts causally efficacious.

By virtue of this filling of the causal gap, the most important demand of intuition – namely that one's conscious efforts have the capacity to affect one's own bodily actions – is beautifully met by the quantum ontology. And in this age of computers, and information, and flashing pixels there is nothing counterintuitive about the ontological idea that nature is built – not out of ponderous classically conceived matter but – out of events, and out of informational waves and signals that create tendencies for these events to occur.

Whitehead deals with the undesirable anthropocentric character of the Copenhagen epistemological position by making the associated human-brain-based quantum objective/subjective events into special cases of a non-anthropocentric general ontology (Whitehead, 1978, pp. 238–239).

Perhaps the main basis for the claim that quantum mechanics is *weird* is the existence of what Einstein called 'spooky action at a distance'. These effects are not only 'spooky' but are also absolutely impossible to achieve within the framework of classical physics. However, if the conception of the physical world is changed from one made out of tiny rock-like entities to a holistic global informational structure that represents tendencies to real events to occur, and in which the choice of which potentiality will be actualized in various places is in the hands of human agents, there is no spookiness about the occurring transfers of information. The postulated global informational structure called the quantum state of the universe is the 'spook' that does the job. But it does so in a completely specified and understandable way, and this renders it basically non-spooky.

In short, the quantum conceptualization is not *intrinsically* counterintuitive, problematic, or weird. It becomes these things only when viewed from a classical perspective that *is counterintuitive* because it denies the causal efficacy of our intentional efforts, is *problematic* because it provides no logical foundation upon which a rational understanding of the occurrence of subjective experience could be built, and is *weird* because it leaves out the mental aspects of nature and chops the body of nature into microscopic, ontologically separate parts that can communicate and interact only with immediate neighbours, thereby robbing both conglomerates and the whole of any possibility of fundamental wholeness or meaningfulness. It is the von Neuman process 1 actions that inject the elements of wholeness and meaning into the quantum universe: without these acts there is nothing but a continuous smear of meaningless un-actualized possibilities (Stapp, 2007a, 2007b).

6.5 Information, God, and values

Information, from the quantum theoretical perspective, is carried by the physical structure that communicates the potentialities created by earlier psycho-physical events to the later ones. This communication of potentialities is an essential part of the process that creates the unfolding and actualization in space and time of the growing sequence of events that constitutes the history of the actual universe. Information resides also in the psychologically described and physically described aspects of these events themselves, and is created by these events.

The information that is created in a *computational process* imbedded in nature resides in the bits that become actualized in this process. This growing collection of bits depends upon the *partitionings* of the quantum smear of possibilities that constitute the universe at some instant (on some space-like surface in the relativistic quantum field theory description) into a set of discrete yes–no possibilities with assigned probabilities. The actualized bits specify the tendencies for future creations of bits. The partitionings specified by the process 1 actions thus lie at the base of the computational notion of information.

But how can these process 1 actions be understood? The partitioning of a continuum into a particular (countable) set of discrete subsets requires a prodigiously powerful choice. This motivates the assumption that the descriptions that *appear to be* continuous within contemporary quantum theory must really be discrete at some underlying level, provided mathematical ideas hold at all at the underlying level.

These processes of choosing are in some ways analogous to the process of choosing the initial boundary conditions and laws of the universe. That is, the free choices made by the human players can be seen as miniature versions of the choices that appear to be needed at the creation of the universe. Quantum theory opens the door to, and indeed demands, the making of these later free choices.

This situation is concordant with the idea of a powerful God that creates the universe and its laws to get things started, but then bequeaths part of this power to beings created in his own image, at least with regard to their power to make physically efficacious decisions on the basis of reasons and evaluations.

I see no way for contemporary science to disprove, or even render highly unlikely, this religious interpretation of quantum theory, or to provide strong evidence in support of an alternative picture of the nature of these 'free choices'. These choices *seem to be* rooted in reasons that are rooted in feelings pertaining to value or worth. Thus it can be argued that quantum theory provides an opening for an idea of nature and of our role within it that is in general accord with certain religious concepts, but that, by contrast, is quite incompatible with the precepts of mechanistic deterministic classical physics. Thus the replacement of classical mechanics by quantum mechanics opens the door to religious possibilities that formerly were rationally excluded.

This conception of nature, in which the consequences of our choices enter not only directly in our immediate neighbourhood but also *indirectly and immediately* in far-flung places, alters the image of the human being relative to the one spawned by classical physics. It changes this image in a way that must tend to reduce a sense of powerlessness, separateness, and isolation, and to enhance the sense of responsibility and of *belonging*. Each person who understands him- or herself in this way, as a spark of the divine, with some small part of the divine power, integrally interwoven into the process of the creation of the psychophysical universe, will be encouraged to participate in the process of plumbing the potentialities of, and shaping the form of, the unfolding quantum reality that it is his or her birthright to help create.

References

Beauregard, M., Schwartz, J., and Stapp, H. P. (2005). Quantum physics in neuroscience and psychology: A neurophysical model of mind-brain interaction. *Philosophical Transactions of the Royal Society of London. Series B, Biological Sciences,* 360(1458): 1309–1327.

Bohr, N. (1958). *Atomic Physics and Human Knowledge.* Reprinted in 1987 as *The Philosophical Writings of Niels Bohr,* vol. II. Woodbridge, CT: Ox Bow Press.

Heisenberg, W. (1958). *Physics and Philosophy: the Revolution in Modern Science.* New York: Harper and Row.

Stapp, H. P. (2005). Quantum interactive dualism: An alternative to materialism. *Journal of Consciousness Studies,* 12(11): 43–59.

Stapp, H. P. (2006). Quantum interactive dualism: The Libet and Einstein–Podolsky–Rosen causal anomalies. *Erkenntnis,* 65(1): 117–142.

Stapp, H. P. (2007a). Quantum approaches to consciousness. In *The Cambridge Handbook for Consciousness*, eds. P. D. Zelazo, M. Moscovitch, and E. Thompson. New York: Cambridge University Press, 881–908.

Stapp, H. P. (2007b). Quantum mechanical theories of consciousness. In *The Blackwell Companion to Consciousness*, eds. M. Velmans and S. Schneider. Oxford: Blackwell Publishing, 300–312.

von Neumann, J. (1955). *Mathematical Foundations of Quantum Mechanics*. Princeton: University Press.

Whitehead, A. N. (1978). *Process and Reality*, corrected ed, eds. D. R. Griffin and D. Sherburne. New York: Macmillan.

PART III
BIOLOGY

7

The concept of information in biology

JOHN MAYNARD SMITH

~

The use of informational terms is widespread in molecular
and developmental biology. The usage dates back to Weismann. In both protein synthesis and in later development,
genes are symbols, in that there is no necessary connection between their form (sequence) and their effects. The
sequence of a gene has been determined by past natural
selection, because of the effects it produces. In biology,
the use of informational terms implies intentionality, in
that both the form of the signal, and the response to it,
have evolved by selection. Where an engineer sees design,
a biologist sees natural selection.

A central idea in contemporary biology is that of information.
Developmental biology can be seen as the study of how information in the genome is translated into adult structure, and
evolutionary biology of how the information came to be there
in the first place. Our excuse for writing a chapter concerning
topics as diverse as the origins of genes, of cells, and of language
is that all are concerned with the storage and transmission of
information.   (Szathmáry and Maynard Smith, 1995)

Let us begin with the notions involved in classical information
theory... These concepts do not apply to DNA because they

Information and the Nature of Reality: From Physics to Metaphysics, eds. Paul Davies
and Niels Henrik Gregersen. Published by Cambridge University Press
© P. Davies and N. Gregersen 2010, 2014.

presuppose a genuine information system, which is composed of a coder, a transmitter, a receiver, a decoder, and an information channel in between. No such components are apparent in a chemical system (Apter and Wolpert, 1965). To describe chemical processes with the help of linguistic metaphors such as "transcription" and "translation" does not alter the chemical nature of these processes. After all, a chemical process is not a signal that carries a message. Furthermore, even if there were such a thing as information transmission between molecules, transmission would be nearly noiseless (that is, substantially nonrandom), so that the concept of probability, central to the theory of information, does not apply to this kind of alleged information transfer. (Mahner and Bunge, 1997)

It is clear from these quotations that there is something to talk about. I shall be concerned only with the use of information concepts in genetics, evolution, and development, and not in neurobiology, which I am not competent to discuss.

7.1 The information analogy

The colloquial use of informational terms is all-pervasive in molecular biology. "Transcription," "translation," "code," "redundancy," "synonymous," "messenger," "editing," "proofreading," "library": these are all technical terms in biology. I am not aware of any confusions arising because their meanings are not understood. In fact, the similarities between their meanings when referring to human communication and genetics are surprisingly close. One example must suffice. In "proofreading," the sequence of the four bases in a newly synthesized DNA strand is compared with the corresponding sequence of the old strand that acted as a template for its synthesis. If there is a "mismatch" (that is, if the base in the new strand is not complementary to that in the old strand according to the pairing rules, A–T and G–C), then it is removed and replaced by the correct base. The similarity of this process to that in which the letters in a copy are compared – in principle, one by one – with those in the original, and corrected if they differ, is obvious. It is also relevant that in describing molecular proofreading, I found it hard to avoid using the words "rule" and "correct."

Molecular biologists, then, do make use of the information analogy in their daily work. Analogies are used in science in two ways. Occasionally, there is a formal isomorphism between two different physical systems. Over 50 years ago, I worked as an aircraft engineer. One thing we wanted to know, in the design stage, was the mode of mechanical vibration of the future airplane. To find out, we built an electrical analog, in which the masses of different parts of the structure were represented by the inductances of coils in the circuit, and elasticity by the capacitances of condensers. The vibrations of the circuit then predicted the vibrations of the aircraft. The justification for this procedure is that the equations describing the electrical and mechanical vibrations are identical. In effect, we had built a special-purpose analog computer. I remember being annoyed, later, to discover that I had been talking prose without knowing it.

Cases of exact isomorphism are rather rare. Much commoner is the recognition of a qualitative similarity, useful in giving insight into an unfamiliar system by comparison with a familiar one. A classic example is Harvey's recognition that the heart is a pump: it is unlikely that he would have had this insight had he not been familiar with the engineering use of pumps. A more controversial example is the fact that both Darwin and Wallace ascribe their idea of evolution by natural selection to a reading of Malthus's *An Essay on the Principle of Population*. A third and more trivial example is that I was led to invent evolutionary game theory by analogy with classical game theory, which analyzes human behavior: as it happens, the main thing I got out of the analogy was a convenient mathematical notation. The point is that scientists need to get their ideas from somewhere. Most often, biologists get them by analogy with current technology, or sometimes with the social sciences. It is therefore natural that during the twentieth century they should have drawn analogies from machines that transduce information. The first deliberate use of such an analogy, by August Weismann, occurred towards the end of the last century, and is described below. Of course, as I will demonstrate, if an analogy is only qualitative, it can mislead as well as illuminate.

However, first I must address the criticisms by Mahner and Bunge quoted at the start of this chapter (Mahner and Bunge, 1997). First, is it true that there is no coder, transmitter, receiver, decoder, or information channel? This sentence does draw attention to some ways in which genetic transcription and translation differ from typical examples of human communication (Figure 7.1).

In the human example, a message is first coded, and then decoded. In the genetic case, although we think of a message in coded form in the mRNA being translated at the ribosome into the amino acid sequence of a protein, it is perhaps odd to think of this as "de"-coding, since it was not "coded" from protein to mRNA in the first place. I do

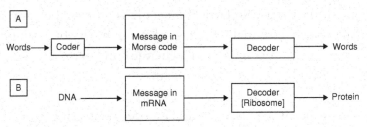

FIG. 7.1 Comparison of transmission of a human message by Morse code (A) and translation of a message coded in DNA into the amino acid sequence of a protein (B).

not think this destroys the analogy between the genetic case and the second part of the human sequence. But it does raise a hard question. If there is "information" in DNA, copied to RNA, how did it get there? Is there any analogy between the origins of the information in DNA and in Morse code? Perhaps there is. In human speech, the first "coder" is the person who converts a meaning into a string of phonemes, later converted to Morse code. In biology, the coder is natural selection. This parallel may seem far-fetched, or even false, to a non-Darwinist. But it is natural selection that, in the past, produced the sequence of bases out of many possible sequences that, via the information channel just described, specifies a protein that has a "meaning," in the sense of functioning in a way that favors the survival of the organism. Where an engineer sees design, a biologist sees natural selection.

What of the claim that a chemical process is not a signal that carries a message? Why not? If a message can be carried by a sound wave, an electromagnetic wave, or a fluctuating current in a wire, why not by a set of chemical molecules? A major insight of information theory is that the same information can be transmitted by different physical carriers. So far, engineers have not used chemical carriers, essentially because of the difficulty of getting information into and out of a chemical medium. The living world has solved this problem.

Finally, what of the objection that the concept of probability is central to information theory, but missing in biological applications? One could as well argue that information cannot be transmitted by the printed word, because typesetting is virtually noiseless. In information theory, Shannon's (1948) measure of quantity of information, $\Sigma p \log p$, is a measure of the *capacity* of a channel to transmit information, given by the number of different messages that could have been sent. The probabilistic aspects of Shannon's theory have been used in neurobiology, but rarely in genetics, because we can get most of what we need from an assumption of *equi-probability*. Given a string of n symbols, each of which can be any one of four equally likely alternatives, Shannon's measure gives $2n$ bits of information. In the genetic message, there are four alternative bases. If they were equally likely, and if each symbol was independent of its neighbors, the quantity of information would be two bits per base. In fact, the bases are not equally likely, and there are correlations between neighbors, so there is some reduction in quantity of information, but it is not very great, and is usually ignored: a greater reduction results from the redundancy of the code. In brief we do not bother with Shannon's measure, because two bits per base is near enough, but we could if we wanted to. As it happens, Gatlin (1972) wrote a whole book applying Shannon's measure to the genetic message. I am not sure that much came from her approach, but at least it shows that the concept of probability does apply to the genetic code. There is a formal isomorphism, not merely a qualitative analogy.

There are difficulties in applying information theory in genetics. They arise principally not in the transmission of information but in its meaning. This difficulty is not peculiar to genetics. In the early days, it was customary to assert that the theory was not concerned with meaning, but only with quantity of information: as Weaver (Shannon and Weaver, 1949) put it: "This word 'information' in communication theory relates not so much to what you do say, as to what you could say." In biology, the question is: How does genetic information specify form and function?

I now describe five attempts, varyingly successful, to apply concepts of information in biology, ending with the problem of biological form. Then, in the concluding section, I use the analogy between evolution and engineering design by genetic algorithms to suggest how ideas drawn from information theory can be applied in biology.

7.2 Weismann and the non-inheritance of acquired characters

Weismann's assertion that acquired characters are not inherited is one of the decisive moments in the history of evolutionary biology. Darwin himself believed in "the effects of use and disuse." What led Weismann to such a counterintuitive notion? Until I happened, rather by chance, to read *The Evolution Theory* (Weismann, 1904), I thought that his reasons were, first, that the germ line is segregated early from the soma and, second, that if you cut the tails off mice, their offspring have normal tails. I thought these were poor reasons. There is no segregation of germ line and soma in plants, yet they are no more likely than animals to transmit acquired characters; and in any case all the material and energy for the growth of the germ cells comes via the soma, so what prevents the soma from affecting the germ cells? As to the mouse tails, this is not the kind of acquired character that one would expect to be transmitted.

I had, of course, done Weismann an injustice. There are two long chapters in *The Evolution Theory* devoted to the non-inheritance of acquired characters. The one argument not used in these chapters is the segregation of the germ line: this was important to Weismann for other reasons. His main argument is that there are many traits that are manifestly adaptive, but that could not have evolved by Lamarckian means, because they could not have arisen as individual adaptations in the first place: an example is the form of an insect's cuticle, which is hardened before it is used, and which therefore cannot adapt during an individual lifetime. It follows that adaptations can evolve without Lamarckian inheritance. But this does not prove that acquired characters are not inherited. His ultimate reason for thinking that they are not was that he could not conceive of a mechanism whereby it could happen. Suppose a blacksmith does develop big arm muscles. How could this influence the growth of his sperm cells, in such a way as to alter the development of an egg fertilized by the sperm, so that the blacksmith's son develops big muscles?

Explaining why he could not imagine such a mechanism, he wrote that the transmission of an acquired character "is very like supposing that an English telegram to China is there received in the Chinese language" (in fact, he uses the telegram analogy twice, in slightly different words). This is remarkable for several reasons. He recognizes that heredity is concerned with the transmission of information, not just of matter or energy. Second, he draws an analogy with a specific information-transducing channel, the telegram. Third, although his insight has been of profound importance for biology, his argument is in a sense fallacious. After all, if a sperm can affect the size of a muscle, why cannot a muscle affect a sperm? In fact, most of the information-transducing machines we use, such as telephones and tape recorders, transmit both ways; they would not be much use if they did not. But some resemble the genetic system in that they transmit only one way. A CD player converts patterns on a disc into sound, but one cannot produce a new disc by singing at the player. I think that the non-inheritance of acquired characters is a contingent fact, usually but not always true – not a logical necessity. Insofar as it is true, it follows from the "central dogma" of molecular biology, which asserts that information travels from nucleic acids to proteins, but not from proteins to nucleic acids.

What, then, of the tails of the mice? Weismann tells us that, when he first spoke of his idea to a zoological meeting in Germany, people replied, "But this must be wrong: everyone knows that, if the tail of a bitch is docked, her puppies have distorted tails" – an interesting example of what Haldane once called Aunt Jobiska's theorem: "It is a fact the whole world knows." The mouse experiment was performed to refute this objection.

A failure to see that heredity is concerned with information, and that information transfer is often irreversible, has unfortunate consequences, as I know to my cost. As a young man, I was a Marxist and a member of the communist party. This is not something I am proud of, but it is relevant. Philosophically, Marxism is unsympathetic to the notion of a gene that influences development, but is itself unaffected: it is undialectical. I do not suggest that the only reason for Lysenko's views was his Marxism – he had less honorable motives – but I think Marxism must take some of the blame. Certainly, it made me uncomfortable with Weismann's views. I spent some 6 months carrying out an experiment to test them. The ability of an adult *Drosophila* to withstand high temperatures depends on the temperature at which the egg was incubated. Not surprisingly, I found that the adaptation is not inherited. For me, the exercise was perhaps not a total waste of time.

7.3 The genetic code

The analogy between the genetic code and human-designed codes such as Morse code or the ASCII code is too close to require justification. But there are some features that are worth noting:

(1) The correspondence between a particular triplet and the amino acid it codes for is arbitrary. Although decoding necessarily depends on chemistry, the decoding machinery (tRNAs, assignment enzymes) could be altered so as to alter the assignments. Indeed, mutations occur that are lethal because they alter the assignments. In this sense the code is symbolic – a point I return to later.

(2) The genetic code is unusual in that it codes for its own translating machinery.

(3) The scientists who discovered the nature of the code, and of the translating machinery, had the coding

analogy constantly in mind, as the vocabulary they used to describe their discoveries makes clear. Occasionally, they were misled by the analogy. An example is the belief that the code would be solved as Linear B was deciphered – by discovering the Rosetta stone. What was needed was a protein of known amino acid sequence, specified by a gene of known base sequence. In fact, the code was not decoded that way. Instead, it was decoded using a "translating machine" – a piece of cell machinery that, provided with a piece of RNA of known sequence, would synthesize a peptide with a sequence that could be determined. But despite such false trails, the information analogy did lead to the solution. If, instead, the problem had been treated as one of the chemistry of protein–RNA interactions, we might still be waiting for an answer.

In an article I came across only when this chapter was almost completed, Sarkar (1996) describes in some detail the history of the idea of a "comma-free code" (Crick, Griffith, and Orgel, 1957). I agree with him that this proved to be a red herring, although I have suggested elsewhere (Maynard Smith, 1999) that it was one of the cleverest ideas in the history of science that turned out to be wrong. But it *was* wrong. It illustrates nicely the fact that analogies in science can be misleading as well as illuminating. But I think that Sarkar is over-eager to point to the failures of the information analogy and to play down its successes. For example, he does not explain that the discovery (Crick et al., 1961) of the relationship between DNA and protein – as a triplet code in which the correct "reading frame" is maintained by accurately counting off in threes, and in which meaning can be destroyed by a "frameshift" mutation – also arose from the coding analogy. It is intriguing that Francis Crick was one of the authors of both papers. As a second example, Sarkar's argument that the code does not enable one to predict amino acid sequences (because of complications such as introns, variations from the universal code, etc.) is seriously misleading; biologists do it all the time.

(4) It is possible to imagine the evolution of complex, adapted organisms without a genetic code. Godfrey-Smith (2000) imagines a world in which proteins play the same central role that they play in our world, but in which their amino acid sequence is replicated without coding. In brief, he suggests that proteins could act as templates for themselves, using 20 "connector" molecules, each with two similar ends, one binding to an amino acid in the template, and another to a similar amino acid in a newly synthesized strand. In such a system, there would be no "code" connecting one set of molecules to another set of chemically different molecules. I agree that such a world is conceivable, and that it lacks a code. I will argue below, however, that the notion of information, and the distinction between genetic and environmental causes in development, would be as relevant in Godfrey-Smith's world as it is in the real world.

7.4 Symbol and "gratuity"

Jacques Monod's (1971) *Chance and Necessity* did not get a good press from philosophers, particularly in the

Anglo-Saxon world. But it contained at least one profound idea: that of *gratuité* (translated, not happily, as gratuity). Jacob and Monod (1961) had discovered how a gene can be regulated. In effect, a "repressor" protein, made by a second "regulatory" gene, binds to the gene and switches it off. The gene can then be switched on by an "inducer," usually a small molecule: lactose for this particular gene. What happens is that the inducer binds to the regulatory protein, and alters its shape, so that the protein no longer binds to the gene and represses it. The point Monod emphasizes is that the region of the regulatory protein to which the inducer binds is different from the region of the protein that binds to the gene; the inducer has its effect by altering the shape of the protein. The result is that, in principle, any "inducer" molecule could switch on, or off, any gene. Of course, all the reactions obey the laws of chemistry, as they must, but there is no chemical necessity about which inducers regulate which genes. It is this arbitrary nature of molecular biology that Monod calls "gratuity."

I think it may be more illuminating to express Monod's insight by saying that, in molecular biology, inducers and repressors are "symbolic": in the terminology of semiotics, there is no necessary connection between their form (chemical composition) and meaning (genes switched on or off). Other features of molecular biology are symbolic in this sense: for example, CAC codes for histidine, but there is no chemical reason why it should not code for glycine. (In passing, I have found the semiotic distinction between symbol, icon, and index illuminating also in animal communication (Maynard Smith and Harper, 1995).)

Sarkar (1996) has an interesting discussion of Monod's notion of gratuity. He interprets Monod as arguing that "The cybernetic account of gene regulation is of more explanatory value than a purely physicalist alternative," but says that this opinion is justified only if cases of gene regulation other than the lactose operon studied by Monod turn out to be of a similar nature. He concludes that "Attempts to generalise the operon model to eukaryotic gene regulation have so far shown no trace of success." I think it would be hard to find a developmental geneticist who would agree with him. As I explain below, Monod's ideas are basic to research in the field.

Linguists would argue that only a symbolic language can convey an indefinitely large number of meanings. I think that it is the symbolic nature of molecular biology that makes possible an indefinitely large number of biological forms. I return to the problem of form later, but first I describe a story of how the information analogy led me up a blind alley, but at the same time prepared me for current discoveries in developmental genetics.

7.5 The quantification of evolution

Around 1960, I conceived the idea that, using information theory, one could quantify evolution simultaneously at three levels: genetic, selective, and morphological. The genetic aspect is easy: the channel capacity is, approximately, two bits per base. Things are complicated by the presence of large quantities of repetitive DNA, but this can be allowed for. The selective level is tricky, but not hopeless. Suppose one asks, "How much selection is needed to program an initially random sequence?" If, reasonably, the selective removal of half the population is regarded as adding one bit of information, then two bits of selection are needed to program each base. The snag is that evolution does not start from a random sequence. Instead, an already programmed gene (or set of genes) is duplicated, and then one copy is altered by selection. However, one can still make a crude estimate of how much selection, measured in bits, is needed to program an existing genome. Kimura (1961), using Haldane's (1957) idea of the "cost of selection," gave a more elegant account of how natural selection accumulates genetic information in the genome.

The hard step is to quantify morphology, but before tackling that question, I want to suggest that the quantification of genetic and selective information in the same units has one, perhaps trivial, use. Occasionally someone, often a mathematician, will announce that there has not been time since the origin of the Earth for natural selection to produce the astonishing diversity and complexity we see. The odd thing about these assertions is that, although they sound quantitative, they never tell us by how much the time would have to be increased: twice as much, or a million times, or what? The only way I know to give a quantitative answer is to point out that if one estimates, however roughly, the quantity of information in the genome, and the quantity that could have been programmed by selection in 5000 MY, there has been plenty of time. If, remembering that for most of the time our ancestors were microbes, we allow an average of 20 generations a year, there has been time for selection to program the genome ten times over. But this assumes that the genome contains enough information to specify the form of the adult. This is a reasonable assumption, because it is hard to see where else the information is coming from.

How much information is needed to specify the form of the adult? Clearly, one does not have to specify the nature and position of every atom in the body, because not everything is specified. This suggested that one asks how much information is required to specify those features shared by two individuals of the same genotype – for example, monovular twins. For simplicity, imagine a pair of two-dimensional organisms (it is easy to extend the argument to three dimensions). Form an image of each as a matrix of black and white dots (in effect, pixels: again, one can extend the argument to more than two kinds of pixel). Start with minute pixels: then identical twins will differ. Gradually enlarge the pixels, until the images of identical twins are the same. Then the information required equals the number of pixels in the image.

It is only necessary to describe the method to see what is wrong with it. Imagine three black-and-white pictures: the first a pattern of random dots, the second the *Mona Lisa*, and the third a black circle on a white ground. The first would indeed require a quantity of information equal to the number of pixels. The *Mona Lisa* could be described in fewer bits, because of the correlations between neighboring dots, but would still require a lot of information. The circle could be specified by saying, if $(x-a)^2 + (y-b)^2 < r^2$, then black, else white (where ab is the center of the circle, and r its radius). One might argue that this is irrelevant, because genes do not know about coordinate geometry, but this would be a mistake. Most simple forms – and a circle is an example – can be generated by simple physical processes, so that all the genome need do is to specify a few physical parameters: for example, reaction rates can be fixed by specifying enzymes.

The fallacy of the "pixel" line of approach is that the genome is not a description of the adult form but a set of instructions on how to make it: it is a recipe, not a blueprint.

7.6 Is the genome a developmental program?

There is, I think, no serious objection to speaking of a genetic code, or to asserting that a gene codes for the sequence of amino acids in a protein. Certainly, a gene requires the translating machinery of a cell – ribosomes, tRNAs, etc. – but this does not invalidate the analogy: a computer program needs a computer before it can do anything. For an evolutionary biologist, the point is that the translating machinery can remain constant in a lineage (although it needs an unchanging genetic program to specify it), yet changes in the genetic program can lead to changes in proteins.

One objection could be that a gene specifies only the amino acid sequence of a protein, but not its three-dimensional folded shape. In most cases, given appropriate physical and chemical conditions, the linear string of amino acids will fold itself up. Folding is a complex dynamic process: it is not yet possible to predict the three-dimensional structure from the sequence. But the laws of chemistry and physics do not have to be coded for by the genes: they are given and constant. In evolution, changes in genes can cause changes in proteins, while the laws of chemistry remain unchanged.

However, an organism is more than a bag of specific proteins. Development requires that different proteins be made at different times, in different places. A revolution is now taking place in our understanding of this process. The picture that is emerging is one of a complex hierarchy of genes regulating the activity of other genes. Today, the notion of genes sending signals to other genes is as central as the notion of a genetic code was 40 years ago.

First, an experiment (Halder, Callaerts, and Gehring, 1995). There is a gene, *eyeless* (also known as *ptx3*), in the mouse. Mutations in this gene (in homozygotes) cause the mouse to develop without eyes, suggesting that the unmutated form of the gene plays some role in eye development. The normal mouse gene has been transferred to the fruit fly, *Drosophila*, and activated at various sites in the developing fly (Halder, Callaerts, and Gehring, 1995). If it is activated in a developing leg, then an eye develops at the site: not, of course, a mouse eye, but a compound fly eye. This suggests that the gene is sending a signal, "make an eye here"; more precisely, it is locally switching on other genes concerned with eye development.

Why should a mouse gene work in a fly? Presumably, the common ancestor of mouse and fly, some 500 million years ago, had the ancestor of the gene: this is confirmed by the presence in *Drosophila* of a gene with a base sequence very similar to the mouse *eyeless* gene. What was the gene doing in that remote ancestor? We do not know, but a plausible guess is that the ancestor had a pair of sense organs on its head – perhaps one or a small cluster of light-sensitive cells – and that the differentiation of these cells, from undifferentiated epidermal cells, was triggered by the ancestral gene.

This raises questions about the nature of the signals that are passing. I argued above that the inducers and repressors of gene activity are symbolic, in the sense that there is no necessary chemical connection between the nature of an inducer and its effects. In Jacob and Monod's original experiments, genes metabolizing the sugar lactose were switched on by the presence of lactose in the medium. This is obviously adaptive; there would be no point in switching on the genes if there was nothing for them to do. But if it was selectively advantageous for these genes to be switched on by a different sugar – say maltose – then changes in the regulatory genes that brought this about would no doubt have evolved.

Yet the experiment described above suggests that the gene responsible for initiating eye development has been conserved for 500 million years. If genes are symbolic, why should this be so? Words are symbols, and are not conserved. The words used to describe a given object change, so why has not the gene used to elicit an eye changed? The question is made more acute by the fact that signaling genes do sometimes acquire new meanings. In evolution, it often happens that a regulatory gene is duplicated: one copy retains its original function, and the other changes slightly, and acquires a new function. I think that the extreme conservatism of many signaling genes can be explained as follows. Regulatory genes are often arranged hierarchically: gene A controls genes B, C, D ... and each of B, C, and D controls yet other genes. Adaptive evolutionary changes are likely to be gradual, and this rules out changes in the initial gene in a regulatory hierarchy. The gene *eyeless*, specifying where an eye is to develop, is likely to be such an initial gene, and so has been conserved. But the point I want to make here is that it is hard even to think about the problem if one does not think of genes sending signals, and if one does not recognize that the signals are symbolic.

To date, then, there is talk of genes "signaling" to other genes, of the genome "programming" development, and so on. Informational terminology is invading developmental biology, as it earlier invaded molecular biology. In the next section I try to justify this usage.

7.7 Evolution theory and the concept of information in biology

I start with a concept of information that has the virtue of clarity, but that would rule out the current usage of the concept in biology. Dretske (1981) argues as follows. If some variable, A, is correlated with a second variable, B, then we can say that B carries information about A; for example, if the occurrence of rain (A) is correlated with a particular type of cloud (B), then the type of cloud tells us whether it will rain. Such correlations depend on the laws of physics, and on local conditions, which Dretske calls "channel conditions."

With this definition, there is no difficulty in saying that a gene carries information about adult form; an individual with the gene for achondroplasia will have short arms and legs. But we can equally well say that a baby's environment carries information about its growth; if it is malnourished, it will be underweight. Colloquially, this is fine; a child's environment does indeed predict its future. But biologists draw a distinction between two types of causal chain – genetic and environmental, or "nature" and

"nurture" – for a number of reasons. Differences due to
nature are likely to be inherited, whereas those due to nurture are not; evolutionary changes are changes in nature,
not nurture; traits that adapt an organism to its environment are likely to be due to nature. For these reasons,
the nature–nurture distinction has become fundamental
in biology. Of course, the distinction could be drawn
without using the concept of information, or applying it
specifically to genetic causes. However, as the examples
discussed above demonstrate, informational language has
been used to characterize genetic, as opposed to environmental, causes. I want now to try to justify this usage.

I will argue that the distinction can be justified only
if the concept of information is used in biology only for
causes that have the property of intentionality (Dennett,
1987). In biology, the statement that A carries information
about B implies that A has the form it does because it
carries that information. A DNA molecule has a particular
sequence because it specifies a particular protein, but a
cloud is not black because it predicts rain. This element
of intentionality comes from natural selection.

I start with an engineering analogy. An engineer interested in genetic algorithms wants to devise a program to
play a competitive game. For simplicity, he chooses Fox
and Geese, a game played on a draughts board in which
four "geese" try to corner a "fox." (As it happens, I played
with the "evolution" of a program to play this game as long
ago as the 1940s. Without a computer, I could not tackle
more difficult games, but Fox and Geese proved easily sol-
uble.) The engineer first invents a number of "rules" for
the geese (for example, keep in line, do not leave gaps, keep
opposite the fox). Each rule has one or more parameters

(for example, for the gap rule, specifying the position of any gaps). He then arranges for a bit string to specify these parameters, and the weightings to be given to the different rules when selecting the next move. He then performs a typical genetic algorithm experiment, starting with a population of random strings, allowing each to play against an efficient fox, selecting the most successful, and generating a new population of strings, with random mutation. For a simple game like Fox and Geese, he will finish up with a program that wins against any fox strategy; things are a bit harder for chess. This procedure is illustrated in panel A of Figure 7.2.

If, instead of using a genetic algorithm approach, the engineer had simply written an appropriate program, no one, I think, would object to saying that the program carried information, or at least instructions, embodying his intentions. By analogy, I want to say that, in the process illustrated in panel A of Figure 7.2, there is information in the bit string, which has been programmed by selection, and not by the engineer. This usage is justified by the fact that, presented with a bit string and the moves that it generated, it would be impossible to tell whether it had been designed by the engineer directly, or by selection between genetic algorithms.

Biological evolution is illustrated in panel B of Figure 7.2. It differs from panel A in two ways. First, a coding stage is present. Second, selection based on success in the game is replaced by survival and reproduction ("fitness") in a specific environment. I do not think the latter difference is important.

I think that the analogy between panels A and B justifies biologists in saying: that DNA contains information

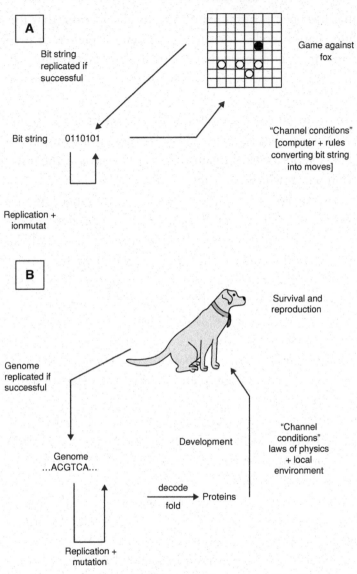

FIG. 7.2 Comparison of selection of a "genetic algorithm" to play a game of Fox and Geese (A) and biological evolution (B).

that has been programmed by natural selection; that this information codes for the amino acid sequence of proteins; that, in a sense that is much less well understood, the DNA and proteins carry instructions, or a program, for the development of the organism; that natural selection of organisms alters the information in the genome; and finally, that genomic information is "meaningful" in that it generates an organism able to survive in the environment in which selection has acted.

The weakness of these models, both engineering and biological, is that they do not tell us where the "rules" come from. In the engineering case, the success of the procedure depends on the ingenuity with which the rules were chosen. In the biological case, the rules depend on the laws of physics and chemistry; organisms do not have to invent, or evolve, rules to tell a string of amino acids how to fold up. But there are higher-level rules, depending on the following facts: that cells divide repeatedly; that every cell contains a complete genome; that cells can signal to their neighbors; that genes can be switched on or off by other genes; and that states of gene switching can be transmitted through cell division to daughter cells. Research in developmental biology is concerned with identifying regulatory genes, and with identifying the higher-level rules with parameters that the genes control.

It should now be clear why biologists wish to distinguish between genetic and environmental causes. The environment is represented in panel B of Figure 7.2 by the "channel conditions." The laws of physics do not change, but the local environment might do. Fluctuations in the environment are a source of noise in the system, not of information. Sometimes, organisms do adapt to changes in the environment during their lifetime, without genetic evolution. For example, pigment develops in the skin of humans exposed to strong sunlight, protecting against ultraviolet rays. Such adaptive responses require that the genome has evolved under natural selection to cope with a varying environment. What is inherited is not the dark pigment itself, but the genetic mechanism causing it to appear in response to sunlight.

This has been a natural history of the concept of information in biology, rather than a philosophical analysis. The concept played a central role in the growth of molecular genetics. The image of development that is emerging is one of a complex hierarchy of regulatory genes, and of a signaling system that is essentially symbolic. Such a system depends on genetic information, but the way in which that information is responsible for biological form is so different from the way in which a computer program works that the analogy between them has not, I think, been particularly helpful – although it is a lot nearer the truth than the idea that complex dynamic systems will generate biological forms "for free." A less familiar idea that has been central both to molecular biology and to development is Monod's notion of "gratuity," which I think is most clearly expressed by saying that molecular signals in biology are symbolic.

Given the central role that ideas drawn from a study of human communication have played, and continue to play, in biology, it is strange that so little attention has been paid to them by philosophers of biology. I think it is a topic that would reward serious study.

7.8 Conclusions

In colloquial speech, the word "information" is used in two different contexts. It may be used without semantic implications; for example, we may say that the form of a cloud provides information about whether it will rain. In such cases, no one would think that the cloud had the shape it did because it provided information. By contrast, a weather forecast contains information about whether it will rain, and it has the form it does because it conveys that information. The difference can be expressed by saying that the forecast has intentionality (Dennett, 1987), whereas the cloud does not. The notion of information as it is used in biology is of the former kind; it implies intentionality. It is for this reason that we speak of genes carrying information during development, and of environmental fluctuations not doing so.

A gene can be said to carry information, but what of a protein coded for by that gene? I think one must distinguish between two cases. A protein might have a function directly determined by its structure – for example, it may be a specific enzyme, or a contractile fiber. Alternatively, it might have a regulatory function, switching other genes on or off. Such regulatory functions are arbitrary, or symbolic. They depend on specific receptor DNA sequences, which have themselves evolved by natural selection. The activity of an enzyme depends on the laws of chemistry and on the chemical environment (for example, the presence of a suitable substrate), but there is no structure that can be thought of as an evolved "receiver" of a "message" from the enzyme. By contrast, the effect of a regulatory protein does depend on an evolved receiver of the information it carries: the *eyeless* gene signals "make an eye here," but only because the genes concerned with making an eye have an appropriate receptor sequence. In the same way, the effect of a gene depends on the cell's translating machinery: ribosomes, tRNAs, and assignment enzymes. For these reasons, I want to say that genes and regulatory proteins carry information, but enzymes do not.

A very similar conclusion about the concept of information in biology has been reached by Sterelny and Griffiths (1999). In particular, they write: "Intentional information seems like a better candidate for the sense in which genes carry developmental information and nothing else does." Justifying this view, they add, "A distinctive test of intentional or semantic information is that talk of error or misrepresentation makes sense." In biology, misrepresentation is possible because there is both an evolved structure carrying the information, and an evolved structure that receives it.

In human communication, the form of a message depends on an intelligent human agent; forecasts are written by humans (or by computers that were programmed by humans), and are intended to alter the behavior of people who read them. There are intelligent senders and receivers. How, then, can a genome be said to have intentionality? I have argued that the genome is as it is because of millions of years of selection, favoring those genomes that cause the development of organisms to be able to survive in a given environment. As a result, the genome has the base sequence it does because it generates an adapted organism. It is in this sense that genomes have intentionality. Intelligent design and natural selection produce similar results. One justification for this view is that programs designed by humans to produce a result are similar to, and may be indistinguishable from, programs generated by mindless selection.

References

Apter, M. J., and Wolpert, L. (1965). Cybernetics and development I. Information theory. *Journal of Theoretical Biology*, 8: 244–257.

Crick, F. H. C., Griffith, J. S., and Orgel, L. E. (1957). Codes without commas. *Proceedings of the National Academy of Sciences of the United States of America*, 43: 416–421.

Crick, F. H. C., Barnett, L., Brenner, S., and Watts-Tobin, R. J. (1961). General nature of the genetic code. *Nature*, 192: 1227–1232.

Dennett, D. (1987). *The Intentional Stance*. Cambridge, MA: MIT Press.

Dretske, F. (1981). *Knowledge and the Flow of Information*. Cambridge, MA: MIT Press.

Gatlin, L. L. (1972). *Information Theory and the Living System*. New York: Columbia University Press.

Godfrey-Smith, P. (2000). On the theoretical role of "genetic coding." *Philosophy of Science*, 67: 26–44.

Haldane, J. B. S. (1957). The cost of natural selection. *Journal of Genetics*, 55: 511–524.

Halder, G., Callaerts, P., and Gehring, W. J. (1995). Induction of ectopic eyes by targeted expression of the *eyeless* gene in *Drosophila*. *Science*, 267: 1758–1791.

Jacob, F., and Monod, J. (1961). On the regulation of gene activity. *Cold Spring Harbor Symposia on Quantitative Biology*, 26: 193–211.

Kimura, M. (1961). Natural selection as a process of accumulating genetic information in adaptive evolution. *Genetical Research, Cambridge*, 2: 127–140.

Mahner, M., and Bunge, M. (1997). *Foundations of Biophilosophy*. Berlin, Heidelberg, New York: Springer-Verlag.

Maynard Smith, J. (1999). Too good to be true. *Nature*, 400: 223.

Maynard Smith, J., and Harper, D. G. C. (1995). Animal signals: Models and terminology. *Journal of Theoretical Biology*, 177: 305–311.

Monod, J. (1971). *Chance and Necessity*. New York: Knopf.

Sarkar, S. (1996). Biological information: A skeptical look at some central dogmas of molecular biology. In *The Philosophy and History of Molecular Biology*, ed. S. Sarkar. Dordrecht: Kluwer Academic Publishers, 157–231.

Shannon, C. E. (1948). A mathematical theory of communication. *Bell System Technical Journal*, 27: 279–423, 623–656.

Shannon, C. E., and Weaver, W. (1949). *The Mathematical Theory of Communication*. Urbana, IL: University of Illinois Press.

Sterelny, K., and Griffiths, P. E. (1999). *Sex and Death: An Introduction to the Philosophy of Biology*. Chicago, IL: University of Chicago Press.

Szathmáry, E., and Maynard Smith, J. (1995). The major evolutionary transitions. *Nature*, 374: 227–232.

Weismann, A. (1904). *The Evolution Theory*, trans. J. A. and M. R. Thomson. London: Edward Arnold.

8

What is missing from theories
of information?

TERRENCE W. DEACON

~

Theories of information that attempt to sort out problems concerning the status and efficacy of its content – as it is understood in thoughts, meanings, signs, intended actions, and so forth – have so far failed to resolve a crucial dilemma: how what is represented could possibly have physical consequences. The legacy of this has been played out in various skeptical paradigms that either conclude that content is fundamentally relativistic, holistic, and ungrounded, or else is merely epiphenomenal and ineffectual except for its arbitrary correlation with the physical properties of the signs that convey it. In this chapter I argue that the apparent conundrums that make this notion controversial arise because we begin our deliberations with the fallacious assumption that in order for the content of information to have any genuine real-world consequences it must have substantial properties, and so must correspond to something present in some form or other. By contrast, I will show that this assumption is invalid and is the ultimate origin of these absurd skeptical consequences.

Information and the Nature of Reality: From Physics to Metaphysics, eds. Paul Davies and Niels Henrik Gregersen. Published by Cambridge University Press © P. Davies and N. Gregersen 2010, 2014.

The crucial property of content that must be taken into account is exactly the opposite: its absence. But how is it possible for a specific absence to have definite causal consequences? A crucial clue is provided by Claude Shannon's analysis of information in terms of constraint on the entropy (possible variety) of signs/signals (Shannon, 1948; Shannon and Weaver, 1949). In other words, the capacity to convey information is dependent on a relationship to something that is specifically not produced. But such a change in the Shannon entropy of a physical medium is also necessarily a physical change, and this must be a product of extrinsic work. In addition, contra Shannon, even when there is no change in Shannon entropy where a change is possible, this can be informative because it indicates the absence of some specific form of extrinsic influence. Both conditions are determined with respect to a potential influence on the form of the semiotic medium that is extrinsic to it. These explicit extrinsic sources of potential signal constraint constitute the ground of the referential capacity by which information is defined. Moreover, I will argue that a process capable of interpreting the reference implicit in such signal constraints depends on coupling the context-dependent signal generation process to a specific extrinsic source of that constraint that is relevant to the existence and maintenance of this interpretive process. Such a linking relationship is provided by a self-sustaining non-equilibrium thermodynamic process, such as that characterizing living organisms. Such a process is capable of interpreting something as information about something else because such systems are necessarily open to and dependent on a precise correlation between intrinsic dynamics and extrinsic conditions. Thus constraints exhibited in certain substrates, which are in some way cor-related both with maintenance of this unstable condition and with non-intrinsic or absent conditions that are relevant to this maintenance, can "matter" to its persistence. Failure to explain this relationship to absence is why past efforts to reduce information to a measurable physical substrate (Shannon information) or to treat it as a primitive nonphysical phenomenon (versions of phenomenology) both led to absurd consequences.

8.1 Introduction

It is often said that we are living in the "information age," but although we use the concept of information almost daily without confusion, and we build machinery (computers) and network systems to move, analyze, and store it, I believe that we still do not really know what *it* is. The ubiquity of the concept of information in our present age is largely a consequence of the invention, perfection, and widespread use of computers and related devices. In our everyday lives information is a necessity and a commodity. We routinely measure the information capacity of silicon, magnetic, or laser data-storage devices, and find ourselves guarding, sharing, or selling information. The importance of understanding and managing information has penetrated to the most technical and most mundane realms of daily life. Molecular biologists have recently mapped the molecular information "contained" in the human genome, and the melding of computer technology with molecular biology has spawned an entirely new field, dubbed bioinformatics, that promises radical new medical technologies and unprecedented threats to privacy. Even household users of internet communication are sensitive to the information bandwidth of the cable and wireless networks that they depend on for connection to the outside world.

It is my contention, however, that we are currently working with a set of assumptions about information that are just barely sufficient to handle the tracking of its most minimal physical and logical attributes, but which are insufficient to understand either its defining representational character or its pragmatic consequences. For more than half a century we have known how to measure the information-conveying capacity of any given communication medium, and yet we cannot give an account of how this relates to the content that this signal may or may not represent. These are serious shortcomings that impede progress in a broad range of endeavors, from the study of basic biological processes to the analysis of global economics.

It is a disturbing fact that, despite the centrality of the concept in our daily lives, we are entirely lacking a clear physical account that explains how information about some abstract concept can have massive and sometimes devastating physical consequences. Consider the concept of "patriotism." Despite the fact that there is no specific physical object or process that constitutes the content of this word, and nothing intrinsic to the sound of the word or its production by a brain that involves more than a tiny amount of energy, its use can contribute to the release of vast amounts of energy unleashed to destroy life and demolish buildings (as in warfare). This is evidence that we are both woefully ignorant of a fundamental causal principle in the universe and in desperate need of such a theory.

In many ways, we are in a position analogous to the early-nineteenth-century physicists in the heyday of the industrial age (with its explosive development of self-powered machines for transportation, industry, timekeeping, etc.), whose conception of energy was still framed in terms of ethereal substances, such as "caloric," "phlogiston," and the "élan vital" that were presumably transferred from place to place to animate machines and organisms. The colloquial notion of information is likewise conceived of in substance-like terms, as for example when we describe movement, storage, or sales of information. The development of the general concept of energy took many decades to clarify, even though the exploitation of energy was a defining feature of that era. The concept was ultimately demystified by recognizing that energy was not a substance, but rather a constant dynamical parameter that was transformed and yet conserved in processes of induced change. The conceptions of energy as ineffable ether or as special substance were abandoned for a dynamical relational account. With this reframing, many once-mysterious phenomena became subject to exact analysis and the basis for myriad new technologies.

Similarly, I argue that in order to develop a full scientific understanding of information we will be required to give up thinking about it, even metaphorically, as some artifact or commodity. To make sense of the implicit representational function that distinguishes information from other merely physical relationships, we will need to find a precise way to characterize its defining non-intrinsic feature – its referential content – and show how it can be causally efficacious despite its physical absence. The enigmatic status of this relationship was eloquently, if enigmatically,

190

framed by Brentano's use of the term "inexistence" when describing mental phenomena.

Every mental phenomenon is characterized by what the Scholastics of the Middle Ages called the intentional (or mental) inexistence of an object, and what we might call, though not wholly unambiguously, reference to a content, direction toward an object (which is not to be understood here as meaning a thing), or immanent objectivity.

This intentional inexistence is characteristic exclusively of mental phenomena. No physical phenomenon exhibits anything like it. We can, therefore, define mental phenomena by saying that they are those phenomena which contain an object intentionally within themselves.   (Brentano, 1874)

As I will argue below, both the engineer's identification of information with the reduction of signal uncertainty and the intuitively attractive phenomenological conception of information as an irreducible "aboutness" relationship that is "always already there," simply take this enigmatic relationship to something not-quite-existent for granted. The first takes it for granted, but then proceeds to bracket it from consideration to deal with physically measurable features of the informational medium. The second treats it as an unanalyzed primitive, and brackets its necessary physicality and efficacy from consideration in order to focus on intrinsic attributes. Neither characterization provides criteria that explicitly distinguish merely physical or logical relationships from those that convey information.

The concept of information is a central unifying concept in the sciences. It plays crucial roles in physics, computation and control theory, biology, cognitive neuroscience, and of course the social sciences. It is, however, employed somewhat differently in each field, to the extent that the aspects of the concept that are most relevant to each may be almost entirely non-overlapping. More seriously, the most precise and technical definition used in communication engineering, computational theory, and quantum physics completely ignores those features that distinguish information from any other causal relationship. This promiscuity threatens to make the concept of information either so amorphous that it provides no insight into the physical distinctiveness of living and mental relationships, or else licenses a retreat into a sort of methodological dualism.

Ultimately, the concept of information has been a victim of a philosophical impasse that has a long and contentious history: the problem of specifying the ontological status of the representations or contents of our thoughts. The problem that lingers behind definitions of information boils down to a simple question: How can the content (aka meaning, reference, significant aboutness) of a sign or thought have any causal efficacy in the world if it is by definition not intrinsic to whatever physical object or process represents it? In other words, there is a paradox implicit in representational relationships. The content of a sign or signal is not an intrinsic property of whatever physically constitutes it. Rather, exactly the opposite is the case. The property of something that warrants calling something information, in the usual sense, is that it is something that the sign or signal conveying it is not. I will refer to this as "the absent content problem." Classic conundrums about the nature of thought and meaning all trace their origin to this simple and obvious fact.

This relationship has often been framed as a mapping or correspondence between a sign or idea in the mind and this something else, which is not present. As countless critics have pointed out, however, this superficially reasonable account fails to identify any features of this relationship that distinguish it from other, merely physical, relationships. Consider the classic example of a wax impression left by a signet ring in wax. Except for the mind that interprets it, the wax impression is just wax, the ring is just a metallic form, and their conjunction at a time when the wax was still warm and malleable was just a physical event in which one object alters another when they are brought into contact. In these facts there is nothing to distinguish it from any other physical interaction. Something more makes the wax impression a sign that conveys information. It must be interpreted by someone. Unfortunately, this obvious answer is ultimately circular. What we invoke with an interpreting mind is just what we hope to explain. The process we call interpretation is the generation of mental signs interpreting extrinsic signs. So we are left with the same problem inside as outside the mental world. The problem of specifying how a specific content is both not physically present and yet inheres in some way in the sign and interpretive process is no better grounded in neurological processes than it is outside of brains.

8.2 Meanings of information

There is, additionally, a troublesome ambiguity in the term "information" that tends to confuse the analysis. This term is used to talk about a number of different kinds of relationship, and often interchangeably without discerning between them. It can refer to the sign or signal features themselves, irrespective of any meaning or reference, as in the information content in bits (binary digits) of the computer file encoding this chapter. This is sometimes called syntactic information. It can refer to what these bits refer to, as in the ideas I am hoping to communicate. This is sometimes called semantic information. And it can refer to that aspect of these ideas that is news to you the reader, and thus not merely redundant as it might be to experts in the field. This is sometimes called pragmatic information. Currently, the first of these meanings has grown in prominence, mostly as a result of our contemporary familiarity with, and dependence on, computing.

This document was created on multiple computers, and in the process I have shared draft versions back and forth with colleagues by sending the information over the Internet. But what exactly was sent? The answer is: a series of high and low voltage values, crudely analogous to Morse code, organized into sets of eight 1s and 0s that together code for an alpha-numeric character or some operation with respect to these. The computers at each end of the process are set up to encode and decode this sequence of shifts in voltage in the same way. For the user, this is accomplished invisibly. All we see are letters arranged on the screen. Clearly, using the terms above we can say that syntactic information is being exchanged back and forth, in the form of numbers of distinguishable signals, and hopefully these signals also convey semantic and pragmatic information as well.

But would we still call it information in any of these senses if there were no people involved? And to make it a bit more like science fiction, would we still call it information if by wild coincidence a large collection of molecules just spontaneously came together to make two computers organized just this way, sending signals back and forth identical to those I have recently sent?[1] This would certainly not qualify as semantic or pragmatic information. What if the signal was composed of randomly generated gibberish rather than English? Would this exclude it from even being described as syntactic information? Presumably, for any finite length of randomly generated signals a suitable code could be defined that would change it into English text. Does this possibility change anything? Ultimately, there is nothing intrinsic to such a string of signals to distinguish it from encrypted English. Is the potential to be translated into a form that could be assigned a meaning sufficient to make it information? If so, then any signal, from any source, however produced and sent, would qualify as syntactic information.

In current computer technology, the rapidly flipping voltages that constitute the operations of a computer can be translated to and from the pattern of microscopic pits burned into a plastic disk or to and from the magnetically modified pattern of iron particles embedded in the surface of a spinning hard disk. But what if we happened upon a naturally occurring pattern of burn pits or magnetized iron atoms in a rock corresponding exactly to these patterns? Would these constitute the same information? Although they might be described as identical syntactic information, they would not be likely to convey identical semantic information. Can it be information at all if it

---

[1] The absurdity of this happening should also tell us something about the complexity hiding behind the notion of information being used here.

derives from a random process? Actually, yes. The chemical reactions caused by unknown molecules in a water sample being tested for contamination, or the radio waves reaching Earth from an exploding star, are typical of the sorts of signal that scientists are able to use as information. Both the patterns that we deliberately create in order to convey an idea, and those we discover in nature, can convey information. Ultimately, this demonstrates that almost anything can qualify as information in the syntactic sense, because this is only an assessment of the potential to inform.

This most basic notion of information corresponds to the contemporary theory of information, originally dubbed the "mathematical theory of communication" by its discoverer Claude Shannon (1948). As we will see, Shannon's definition is more precise than this, but in essence it shows us why it is possible to treat any physical distinction as *potential* information, whether made by humans or the product of some mindless natural process. It specifies what features of a physical material or process are necessary for it to serve as a medium for transmitting or storing information. So, in this sense, when we use the term "information" to refer to signals transmitted over a wire, or ink marks on paper, or the physical arrangement of objects at a crime scene, we are using a sort of shorthand. Without these physical features there is no information, but we are actually referring to something more than the physical pattern; something not present to which these present features correspond.

Identifying the features of physical processes that are the necessary conditions for something to be able to provide information helps to make sense of the enterprise of the natural sciences. Scientific observation and experiment are directed to the task of probing the patterns of things to discover information about how they came to have the properties they have. And the same physical objects or events can yield new information with each change of interpretive apparatus. This open-endedness is the result of there usually being vastly more information potential in natural phenomena than can ever be interpreted. To a brilliant sleuth equipped with the latest tools for materials analysis and DNA testing, almost everything at a crime scene can become a source of information. But being able to specify what physical properties can potentially serve to provide information does not help us to discern how it is that they can be about something else.

8.3 Locating the information in information processing

Shannon's analysis of the logical properties that determine the information capacity of a physical medium helps make sense of the concept of information in computer theory. In the most general sense, the possibility of computing depends on being able to assign referential value to some feature of a physical process and to map a specific logical operation to some physical manipulation of that feature with respect to others that have also been assigned reference. In this sense, one can even talk about arbitrary mechanical operations (or other physical processes) as potential computers. Just as one could come up with a coding scheme that could interpret an arbitrary sequence of signals as an English sentence, so it is possible to find a mapping between an arbitrary physical process and some symbol manipulation process. Of course some physical processes and mechanical (or electronic, or quantum) devices are better than others for this, especially when we desire flexibility in possible mappings. This mapping relationship – assigning reference – is crucial for distinguishing between computing and other merely physical processes. All physical processes are potential computations, but no physical process is intrinsically a computation.

It is in this most general sense that we can also describe mental processes as computing. And yet missing from this analogy is precisely the mapping relationship that distinguishes thought from computing. There is no separate homunculus external to the computation that assigns reference to the relevant physical differences of neural dynamics. Computation is generally described in terms of a syntactic conception of information, and yet it implicitly presupposes a semantic conception, although it can give no account of it. It is sometimes assumed that this referential mapping can be provided by the equivalent of robotic embodiment, so that the input and output of the computing is grounded in physical-world constraints. But this can also be seen as an extension of the physical mechanism underlying the computation to include causal events "outside" some arbitrarily defined boundary separating the "computing device" and the physical environment.

Describing both physical and mental relationships in computational terms is problematic only if this presupposition of mapping is ignored or assumed to be intrinsic. The result is either eliminative reduction or cryptic panpsychism, respectively. In either case, if any physical event is considered to be a computation and the mind is merely a special-purpose computer, then the mind–body problem dissolves. But there is a troubling implication to this collapse of the concept of information to its syntactic meaning only. In such a uniformly informational universe there is no meaning, purpose, value, or agency. In this informational cosmology, networks of informational causality are still just as blindly mechanical as in any Laplacian universe.

To escape this deflationary view of an information universe blindly mechanistically computing us, many have turned to quantum physics to loosen the bonds of mechanistic determinism, both in discussions of consciousness and in terms of information processes. Quantum events appear postponable until they are observed, and quantum objects can be both independent and correlated (entangled) at the same time. Thus notions of causality and of information about that causality appear to be inextricably linked at this level of scale.

For example, in the dominant (although not the only) interpretation of quantum mechanics, events in the world at the quantum level become real (in the sense of being susceptible to classical analysis) only when they are measured. Before such an intervention, no explicit single state exists: only a field of potentiality. This is exemplified by the famous Schrödinger's cat paradox, in which the prior death of a cat in a closed box is dependent on an as-yet-unmeasured quantum state. In this interpretation it is presumed that neither macroscopic state exists until the quantum event is later measured (that is, by an observer). Measurement information about this quantum state is thus treated as a fundamental causal determinant of the transition from quantum indeterminacy to classical determinism. Similarly, another strange quantum phenomenon – quantum entanglement – shares other features with the correspondence relationship associated with information. In simplified terms, this involves an apparent instantaneous correspondence of measurement consequences between particles that are separated and non-interacting. So one might also argue that it exemplifies a sort of intrinsic representation relationship.

These are counterintuitive phenomena, which challenge our normal conceptions of causality, but do they explain the higher-order senses of information? Unfortunately, they do not actually resolve the paradox of the absent content. These features in the quantum realm (for example, superposition, entanglement) resemble correspondence-mapping relationships. Thus we might be tempted to see this as a referential relationship that is intrinsic to the quantum physical relationship. But physical correlation is not aboutness. Whereas measurements of particles that affect measurements of other particles exist in something like a correspondence relationship, this alone does not make one *about* the other, except to an external observer interpreting it. The aboutness does not exist in the interstices between indeterminate quantum events any more than between the gears of a clock, because it is not an intrinsic feature. So in both classical and quantum computation only the syntactic concept of information is invoked. There is nothing intrinsic to computation processes that distinguish them from other physical processes, and nothing intrinsic to the quantum or classical physical features that are manipulated in computations that make them about other features of the world.

8.4 Is information physical?

There is something correct about the link between information and the fabric of causal processes of the world – whether determinate or intrinsically statistical – but there is something missing too. There is something more that we assume when we describe something as information, and indeed something absent from the physical processes and patterns that we recognize as conveying (though not fully constituting) information.

The search for a link between information and physical causality in general requires that we identify a precise physically measurable correlate of information. This is necessary in order to solve engineering problems that involve information systems and to address scientific issues concerning the assessment of information processes in natural systems. A first solution to these practical challenges was formally provided by a Bell Labs researcher in the 1940s, Claude Shannon (Shannon and Weaver, 1949). His "mathematical theory of communication" demonstrated that the capacity of a communication medium to convey or store information could be precisely measured, and that even informational error-correction can be accomplished without any reference to informational content. This laid the foundation for all of modern information technology, but it also left us with a deflationary theory of information, from which content, reference, and significance are excluded and irrelevant.

Claude Shannon's introduction of a statistical approach to the analysis of signals and their capacity to carry information has stood the test of time with respect to any practical accounting of how much information a given

medium can be expected to store or convey to an interpreter. Unfortunately, because this analysis excluded any
reference to problems of defining content or significance,
it has led to the rather different uses of the term that
we have been struggling to define and which are often a
source of theoretical sleight of hand. By bracketing issues
of reference and significance Shannon was able to provide an unambiguous, interpretation-free measure of what
might be called the information-bearing capacity (as distinguished from information itself). Not only does this
work for human-made communication processes, it also
usefully conveys the potentiality of any physical distinction to provide information, such as might be discovered
by scientific experiment or detective work. But for this
purpose it had to stop short of conveying any sense of
how information could come to be about something. And
there is good reason to have avoided this. For different
interpreters or for different scientific instruments the same
physical distinction can provide information about different things, or can be irrelevant and uninformative. What
something is about and how this relationship is mediated
are explicitly a function of external relations and thus are
not able to be mapped to any intrinsic properties.

Information is by definition something in relation to
something else, but in colloquial use the term can refer
either to what is conveyed or what provides the con-
veyance. If, as in Shannon's sense, it refers only to the
latter, then its aboutness and its significance are assumed
potentialities but are temporarily ignored. The danger of
being inexplicit about this bracketing of interpretive context is that one can treat the sign as though it is intrinsically significant, irrespective of anything else, and thus

end up reducing intentionality to mere physics, or else imagine that physical distinctions are intrinsically informational rather than informational only *post hoc*, that is, when interpreted.

However, although Shannon's conception of information totally ignores the issue of what information is *about*, or even that it is about anything, his analysis nevertheless provides an important clue for dealing with the absent-content problem, specifically by showing that absence could have a function at all. This clue is provided by Shannon's negative characterization of information. Shannon's measure of the potential information conveyed by a given message received via a given communication channel is inseparable from the range of signals that could have been received but were not. More precisely, Shannon information is defined as the amount of uncertainty that is removed with the receipt of a given signal. So to measure information requires comparing the potential variety of signals that could have been transmitted with what was transmitted. Perhaps the most important contribution of this analysis was his recognition that measuring the potential signal variety was mathematically analogous to measuring the entropy of a physical system, such as an ideal gas. Following the advice of the mathematician John von Neumann, he decided to call this variety of possible states the "entropy" of the signal medium (or "channel," as he described it, using the model of a communication channel between a sender and recipient of a message). This decision, like the decision to define information capacity with respect to signal constraint, has led to many confused debates about the physical correlates of information. But these analogies are also important hints for expanding the concept of information so that it can again embrace the very features that had to be excluded from this engineering analysis.

8.5 Two entropies

By defining information in negative terms with respect to the potential variety of what could have occurred, Shannon has inadvertently pointed us toward the relevant physical property of the signal or sign medium that provides access to what it is about, which is also a negative attribute. The key to seeing this link is simply recognizing that the representing medium, whatever form it takes, is necessarily a physical medium. It is something present that is taken to be about something not immediately present. The reduction in signal entropy has the potential to carry information because it reflects the consequences of physical work and thus the openness of this physical signal medium to extrinsic influence. In thermodynamic terms (Boltzmann, 1866), a change in the state of a physical system that would not otherwise occur is inevitably characterized by a local reduction in its physical entropy (I will describe this as "Boltzmann entropy" to distinguish it from Shannon entropy) resulting from work done on that system from outside. According to Shannon, the information-bearing capacity of a signal is proportional to the improbability of its current physical state. But an information medium is a physical medium, and a physical system in an improbable state reflects the effects of prior physical work, which perturbed it from some more probable state or states. In this way, the Shannon information embodied in signal constraints implicitly represents this work. It is in this sense that Shannon entropy is intimately related to Boltz-mann entropy. It is explicitly the change in the Boltzmann entropy of the medium that is the basis of signal reference, because this is necessarily a reflection of some extrinsic influence (Deacon, 2007, 2008).

However, the relationship is more subtle than just the consequence of physical work to change a signal medium. Although this relation to work is fundamental, referential information can be conveyed both by the effect of work and the evidence that no work has been done (Deacon, 2007, 2008). Thus, no news can be news that something anticipated has not yet occurred. This demonstrates that Shannon information and referential information are not equivalent. This is again because the signal constraint is not something located *in* the signal medium: it is rather a relationship between what is and what could have been its state at any given moment. A reduction in variabil-ity is a constraint, and a constraint is in this case not an intrinsic property but a relational property. It is defined with respect to what is not present. So implicitly, a physi-cal system that exhibits constraint is in that configuration because of extrinsic influences – but likewise if the sign medium exhibits no constraint or change from some sta-ble state, it can be inferred that there was no extrinsic influence doing work on it. The relationship of present to absent forms of a sign medium embodies the openness of that medium to extrinsic intervention, whether or not any interaction has occurred. Importantly, this also means that the possibility of change due to work, not its actual effect, is the signal feature on which reference depends. This is what allows absence itself, absence of change, or being in a highly probable state to be informative.

Consider, for example, a typo in a manuscript. It can be considered a reduction of referential information because it reflects a lapse in the constraint imposed by the language that is necessary to convey the intended message, and yet it is also information about the proficiency of the typist, information that might be useful to a prospective employer. Or consider a technician diagnosing the nature of a video hardware problem by observing the way the image has become distorted. What is signal and what is noise is not intrinsic to the sign medium, because this is a determination with respect to reference. But in either case the deviation from a predicted or expected state is taken to refer to an otherwise unobserved cause. Similarly, a sign that does not exhibit the effects of extrinsic influence – for example, setting a burglar alarm to detect motion – can equally well provide information that a possible event (a break-in) did *not* occur. Or consider the thank-you note not sent, or the tax return not submitted on time. Here, even the absence of a communication is a communication that can carry significance and have dire consequences.

In all cases, however, the referential capacity of the informational vehicle is dependent on physical work that has, or could have, altered the state of some medium open to extrinsic modification. This tells us that the link between Shannon entropy and Boltzmann entropy is not mere analogy or formal parallelism. It is the ground of reference.

8.6 Darwinism and interpretation

Up to this point of the analysis it has been assumed that the relationships being described have involved signs and signals, and not merely physical events chosen at random. But in fact, *none* of the criteria specified thus far actually distinguishes events and objects that convey information from those that do not. They are *requirements* for something to be information about something else, but they do not in themselves constitute it. Shannon described the necessary conditions for something to have the potential to convey information: providing a syntactical conception. Even the linkage to physical work, while a necessary requirement for referential capacity, is like Shannon's criterion, only a necessary but not sufficient feature of reference: that is, a semantic conception of information. But of course, not just any alteration of entropy constitutes a reference relationship. Although any physical difference *can* be interpreted as information about something else – whether it is the state of the mud on someone's shoes or the presence and evenness of the microwave background radiation of the universe – this is not an intrinsic feature, but something entirely relative to how it is interpreted. This post-hoc dependency does not diminish the necessity of these attributes. It merely demonstrates that they are not sufficient. And yet, as we have seen, reference depends on the responsiveness of the information medium to physical change.

A physical difference becomes informational when it plays a modulatory role in a dynamic process. The potential to inform is dependent on the Shannon–Boltzmann criteria just discussed, but this potential is actualized only as it influences a specifically structured dynamical process. Although we commonly talk about this as an interpretive process, invoking folk psychological assumptions, this is still for the most part a mere promissory note for a missing theory about what dynamical organization is sufficient to constitute such a process. And this heuristic immediately becomes problematic when we attempt to expand the usage to domains such as molecular biology, in which the presence of a homuncular interpreter cannot be invoked.

A key insight into the conditions for interpretation was provided by Gregory Bateson in an oft-cited aphorism proposed as a characterization of information: Information is "a difference that makes a difference" (Bateson, 1972). It is no coincidence that this phrase also would be an apt description of mechanical or thermodynamic work. Implicit in this phrase is the notion that information can be used to change things. And in this sense it has the potential to control work. So putting this idea about the physical basis of interpretation together with the Boltzmannian criterion for referentiality we have: "a medium that is susceptible to being modified by physical work, which is used to modify the state of some other dynamical system because of that system's sensitivity to changes of this medium, and which is differentially capable of performing work with respect to such a change."

This is a complicated definition, but even so it lacks detail concerning the nature of such a dynamical architecture, and so it is still incomplete in a couple of important respects. These have to do with informational relevance and the notion of function from which the normative aspect of information arises: that is, the pragmatic conception of information. Before trying to address these limitations, however, we need to flesh out the requirements for a system with the capability to differentially perform work with respect to a given signal state. This is because these requirements will ultimately provide grounding for these additional features, and the basis upon which the possibility for specific reference can arise.

It is a simple rule of thermodynamics that to be capable of performing work a system must be in a non-equilibrium state. So, any full explanation of what constitutes an interpretive process must include a central role for non-equilibrium dynamics. But non-equilibrium conditions are inherently transient and self-undermining. For a non-equilibrium process to persist, it must rely on supportive environmental conditions (for example, a source of free energy and raw materials) to make up for this spontaneous degradation. In this sense, like the signal medium, it must be open to influences extrinsic to itself: for example, a larger thermodynamic context to which it is well fitted. Thus the presence of a system maintaining a persistent non-equilibrium state capable of performing work with respect to a source of information entails the presence of environmental conditions that promote it.

This is important for the constitution of an interpretive process for two additional reasons. First, the openness to context of an interpreting system cannot merely be sensitivity, as is the case for an information medium. Persistence of a non-equilibrium thermodynamic system requires a quite specific matching of dynamical organization with extrinsic supportive conditions. In other words, there must be a necessary correspondence in form between system and context. Second, for a dynamical non-equilibrium system to be persistent there must be some maintenance of self-similarity, and thus boundedness, to it. It must have a unit identity in at least a loose sense.

Understanding that a process capable of generating information necessarily involves non-equilibrium dynamics also provides a way to address the normativity issue that is implicit in a pragmatic conception of information. Normativity in its various aspects has been a non-trivial problem for correspondence and mapping theories of reference. Following Bickhard (1998, 2000, 2003), I would argue that the normativity that defines representational error is an emergent property of the relationship of the Shannon–Boltzmann referential relationship with respect to the organization of the non-equilibrium processes that interpret it. This follows because of the intrinsic dependence on specific environmental conditions that are required for such a dynamical system to persist. To the extent that a particular signal interpretation effectively contributes to this end, and thus aids the successful maintenance of this supportive correlation, then that particular interpretive response to that particular state of the signal medium will also persist. Of course, the opposite is also possible; hence the possibility of misinterpretation.

But even in the simplest case this presupposes a non-equilibrium process that is precisely organized with respect to both supportive environmental conditions and to some feature of that environment that tends to correlate with those conditions. The establishment of such a reliable relationship is then the transformation of an incidental physical relationship into an information relationship. It is probably the case that this matching of specific referential relationship with a specific dynamical modification of the capacity to perform work can be achieved spontaneously only by an evolutionary process. Not surprisingly, then, this analysis suggests that the generation of information in the full sense is an emergent property of life. Of course, this does not exclude the infinitely many ways that information can be generated and manipulated, indirectly, with respect to living processes. Yet these, too, must at least indirectly embody these same basic criteria. It is in this sense that both the syntactic (Shannon) and semantic (Shannon–Boltzmann) conceptions of information are ultimately dependent on a pragmatic (Shannon–Boltzmann–Darwin) conception (Deacon 2007, 2008). In this way, the process of evolution, in its most general form, can be understood as the necessary originative source for information. Where there is no evolutionary dynamic there is no information in the full sense of the concept.

8.7 Information evolves

The claim that evolution itself constitutes an information creation process needs to be unpacked in order to complete this analysis. There is yet an additional curious – but in hindsight not unexpected – parallel between the Shannonian determination of information capacity and the evolutionary determination of fittedness. Both involve the relationship between a potential and a realized variety. Natural selection depends on the generation of functionally uncorrelated (aka "random") variety of forms (genotypes and phenotypes) followed by the reduction of this variety due to death or the failure to reproduce. In our analysis of the Shannon–Boltzmann relationship, the referential potential of a signal medium was shown to be a consequence of the way extrinsic factors reduce its potential variety (entropy). In the process of natural selection an analogous role is played by conditions in the environment that favor the reproduction and persistence of some variants and not others. It is in this sense that we feel justified in claiming that the traits that are present in any given generation of organisms are adaptations to (favorably correlated with) certain of those conditions. Metaphorically speaking, they could be said to be "about" those conditions.

There are deep disanalogies, however, that are crucial to explaining why this process generates new information. First, the initial variety is not signal variety, not merely variations of some passive substrate. The variety that is subject to the constraining influence of natural selection involves variation of processes and structures associated with the interpretive system itself. Second, the non-equilibrium dynamics of organisms competing with one another to extract or sequester resources is the source of the work that is the basis for this reduction in population "entropy." And third, the variety that is subject to selection is being generated anew in each generation by virtue of what in Shannonian terms would be considered noise (that is, mutations and recombinations) introduced into the signal (that is, genetic inheritance). Thus what amounts to uncorrelated corruptions of the genetic signal and incidental physical attributes can become information to the extent that they result in variations of the interpretive–adaptive process that happen to embody correlated predictive correspondences between the dynamics of interpretation and the supportive conditions enabling this interpretation. The capability of the Darwinian process to generate new information about organism–environment (and by extension the interpreter–context) interdependency is the ultimate demonstration of the post-hoc nature of information. This evolutionary transformation of noise into information is the ultimate demonstration that what makes something information is not intrinsic to any features of the information-conveying medium itself. It is irreducibly relational and systemic, and at every level of analysis dependent on a relationship to something not present.

8.8 Conclusions

The "intentional inexistence" of the content of a thought, the imagined significance of a coincidental event, the meaning of a reading from a scientific instrument, the portent of the pattern of tea leaves, and so on, really is something that is not there. In this sense the Cartesian-derived notion that the content of mind is without extension, whereas the brain processes that realize this content do have extension, is at least partly correct. But to say that this absent content is extensionless is not quite right. The non-produced signal (that is, reduced entropy) that is the basis for Shannonian informative capacity, the non-present work that was or was not the basis for the reference of this signal, and the interpretive options (organism trait variations) selected in an evolutionary process, all have a definite negative extension in the sense that something specific and explicit is missing. In other words, like the space within a container, these are absences that are useful because of the way what is present can exemplify them.

The nearly universal tendency to attribute intentional phenomena to a disembodied realm is a reflection of this negative defining feature, but the apparent paradoxes this creates with respect to the physical efficacy of informational content is the result of misinterpreting this negative feature as though it is in some way substantial in a separate disembodied realm. The modern shift to abandoning all consideration of intentionality in definitions of information, as the concept has come to be used in the sciences, in order to focus entirely on the material–logical attributes of signal differences, has correspondingly stripped the concept of its distinctive value and has led to a reduction of information relationships to relationships of physical difference. As a result this most common and undeniable feature of our existence is often treated as though it is epiphenomenal. Even the recent efforts to reframe intentionality with respect to its embodiment effectively recapitulate a cryptic form of dualism in terms of a variant of dual aspect theory. But avoiding addressing the "inexistence" problem in these ways guarantees that the real-world efficacy of information remains inexplicable.

Like so many other "hard problems" in philosophy, I believe that this one, too, appears to have been a function of asking the wrong sort of questions. Talking about cognition in terms of the mind–brain – implying a metaphysically primitive identity – or talking about mind as the software of the brain – implying that mental content can be reduced to syntactic relationships embodied in and mapped to neural mechanics – both miss the point. The content that constitutes mind is not *in* the brain, nor is it *embodied* in neuronal processes in bodies interacting with the outside world. It is, in a precisely definable sense, that which determines which variations of neural signaling processes are *not* occurring, and that which will in a roundabout and indirect way help reinforce and perpetuate the patterns of neural activity that are occurring. Informational content distinguishes semiosis from mere physical difference. And it has its influence on worldly events by virtue of the quite precise way that it is *not* present. Attempts to attribute a quasi-substantial quality to information or to reduce it to some specific physical property are not only doomed to incompleteness: they ultimately ignore its most fundamental distinctive characteristic.

References

Bateson, G. (1972). *Steps to an Ecology of Mind: Collected Essays in Anthropology, Psychiatry, Evolution, and Epistemology.* Chicago, IL: University of Chicago Press.

Bickhard, M. H. (1998). Levels of representationality. *Journal of Experimental and Theoretical Artificial Intelligence*, 10(2): 179–215.

Bickhard, M. H. (2000). Autonomy, function, and representation. *Communication and Cognition – Artificial Intelligence*, 17(3–4): 111–131.

Bickhard, M. H. (2003). The biological emergence of representation. In *Emergence and Reduction: Proceedings of the 29th Annual Symposium of the Jean Piaget Society*, eds T. Brown and L. Smith. Hillsdale, NJ: Erlbaum, 105–131.

Boltzmann, L. (1866). The Second Law of Thermodynamics. Reprinted in *Ludwig Boltzmann: Theoretical Physics and Philosophical Problems, Selected Writings*, ed. B. McGuinness, trans. P. Foulkes (1974). Dordrecht: Reidel Publishing Co., 13–32.

Brentano, F. (1874). *Psychology From an Empirical Standpoint.* London: Routledge & Kegan Paul, 88–89.

Deacon, T. (2007). Shannon–Boltzmann–Darwin: Redefining information. Part 1. *Cognitive Semiotics*, 1: 123–148.

Deacon, T. (2008). Shannon–Boltzmann–Darwin: Redefining information. Part 2. *Cognitive Semiotics*, 2: 167–194.

Shannon, C. (1948). A mathematical theory of communication. *Bell System Technical Journal*, 27: 279–423, 623–656.

Shannon, C., and Weaver, W. (1949). *The Mathematical Theory of Communication*. Urbana, IL: University of Illinois Press.

9

Information and communication in living matter

BERND-OLAF KÜPPERS

≈

Ever since the elucidation of the molecular basis of living systems, we have known that all elementary processes of life are governed by information. Thus, information turns out to be a key concept in understanding living matter (Küppers, 1990). More than that: the flow of information at all levels of the living system reveals the properties of communication. This means that the information stored in the genome of the organism is expressed in innumerable feedback loops – a process through which the genetic information is continually re-evaluated by permanent interactions with the physical environment to which it is exposed. In this way, the living organism is built up, step by step, into a hierarchically organized network of unmatched complexity.

The fact that all phenomena of life are based upon information and communication is indeed the principal characteristic of living matter. Without the perpetual exchange of information at all levels of organization, no functional order in the living organism could be sustained. The processes of life would implode into a jumble of chaos if they

Information and the Nature of Reality: From Physics to Metaphysics, eds. Paul Davies and Niels Henrik Gregersen. Published by Cambridge University Press
© P. Davies and N. Gregersen 2010, 2014.

were not perpetually stabilized by information and communication. In this chapter, I should like to consider some of the consequences that follow from this for our philosophical understanding of reality.

9.1 About "information" and "communication"

In daily usage, the terms "information" and "communication" are not always clearly distinguished from each other. Yet, even the etymology of the two words indicates that the reference of the concepts cannot entirely overlap. The term "information" – following closely its Latin root *informare* – denotes primarily the formative, and thus instructive function of a message. By contrast, the word "communication" – derived from the Latin word *communicare* – denotes the process by which the sender and the receiver of information try to reach a common understanding. The subject of this understanding is a common evaluation of the information exchanged between the sender and the receiver. Alongside this, the bare instruction for which the word "information" stands seems like a command that results in the mechanical: that is, the unilateral transfer of the information from sender to receiver, without any aim of achieving a common or mutual understanding of the "meaning" of the information being expressed by its operative function.

Thus, if we wish to approach the concept of communication in living matter in its widest sense, we need to examine the relationship between information on the one hand and mutual or common understanding on the other. At the same time, we shall need to demonstrate that concepts such as "information" and "communication" can meaningfully be applied to natural processes. The latter task would seem to raise fewer difficulties for the concept of information than it does for that of communication.

Information, as suggested above, means primarily "instruction," in the sense of a command or step in a computer program. This in turn has the function of imposing a selective condition on the possible processes that can go on in a system. In precisely this sense, living processes are "instructed" by the information that is contained in encoded form in genes. Expressed in the terms of physics: the genome represents a specific physical boundary condition, a constraint that restricts the set of physically possible processes to those that do actually take place within the organism and are directed towards preservation of the system (Küppers, 1992). Thus, the idea of "instruction by information" has a precise physical meaning, and in this context information can be indeed regarded as an objective property of living matter.

It is a harder task to demonstrate the universality of communication. One tends to assume intuitively that this concept is applicable only to the exchange of information between human beings. This assumption arises from the fact that the idea of a "common" understanding seems to make no sense outside the realm of human consciousness. However, this could be a false premise based upon a narrow use of the concept of understanding. Reaching a common understanding usually means reaching agreement. This in turn implies that one must understand one another in the sense that each party can comprehend what the other party intends to communicate. However, attaining a common understanding does not necessarily presuppose any reflections upon the nature or the subject of the communication process, nor does it imply the question of whether the contents of the communication are true or false. Rather, it requires only a mere exchange of information; that is, a number of messages to be sent in both directions – without, however, either party necessarily being aware of the meaning of what is being communicated.

There is thus a subtle difference in scope between "a reflected understanding" and "reaching a coordinated reaction." If we are for a moment willing to put aside the highly sophisticated forms of human understanding, and to operate with a concept of understanding that encompasses only the objectives of achieving a coordinated reaction, then it becomes easy to see how this concept is applicable to all levels of living matter. We thus have to concede that molecules, cells, bacteria, plants, and animals have the ability to communicate. In this case, "communication" means neither more nor less than the reciprocal harmonization and coordination of processes by means of chemical, acoustic, and optical signals.

9.2 About "understanding"

The foregoing arguments have taken me along a path that some philosophers of science have branded "naïve" naturalism. Their criticism is directed especially at the idea that information can exist as a natural object, independently of human beings: that is to say, outside the myriad ways in which humans communicate. This charge of naturalism is heard from quite diverse philosophical camps. However, all such critics share the conviction that only human language can be a carrier of information, and that the use of linguistic categories in relation to natural phenomena is nothing more than a naturalistic fallacy. For representatives of this philosophical position, any talk of information and communication in the natural sciences – as practiced especially in modern biology – is no more than a metaphor that reveals, ultimately, a sadly uncritical usage of terms such as "language" and "understanding."

Let us examine this controversy more closely and ask once more the question of what we actually understand by the word "understanding." The tautological way in which I express this question indicates that one can easily get into a vicious circle when trying to approach the notion of understanding. This is because it seems generally to be the case that one can only understand something if one has understood some other things. This plausible statement is central to philosophical hermeneutics, the best-known and most influential doctrine of human understanding (Gadamer, 1965).

The hermeneutic thesis, according to which any understanding is bound to some other understanding, obviously refers to the total "network" of human understanding in which any kind of understanding is embedded. In other words: any form of communication presupposes some prior understanding, which provides the necessary basis for a meaningful exchange of information. In fact, there seems to be no information in an absolute sense – not even as a plain syntactic structure – as the mere identification of a sequence of signs as being "information" presupposes a foregoing knowledge of signs and sequences of signs. In short: information exists only in a relative sense – that is, in relation to some other information.

Thus, even if we adopt an information-theoretical point of view, there seems to be no obstacle to the hermeneutic circle, according to which a person can only understand something if he has already understood something else. Nevertheless, this perspective contradicts the intentions of philosophical hermeneutics, which puts a completely different construction upon the hermeneutic circle. Within the framework of this philosophy, the pre-understanding of any kind of human understanding is thought to be rooted in the totality of human existence. And this ontological interpretation is intended to lead not to a relative but to an absolute and true understanding of the world.

Moreover, because we use language to comprehend the world, the hermeneutic school regards language as the gate that opens for us the access to our existence. The philosopher Hans-Georg Gadamer (1965, p. 450) has expressed this in the often-quoted sentence: "Being that can be understood is language." Even though some prominent philosophers of the hermeneutic school assign a special role to dialogue, their concept of understanding still fails to possess the objectiveness and relativity that characterize a critical comprehension of human understanding. On the contrary: a world view that rests its claims to validity and truth exclusively upon the rootedness of understanding in human existence has moved to the forefront and become the absolute norm for any understanding at all.

So, in contrast to the relativistic world picture offered to us by modern science, philosophical hermeneutics seeks to propagate a fundamentalism of understanding that is centered firmly on the philosophical tradition of absolute understanding. Moreover, if human language is considered to be a prerequisite for all understanding, human language becomes the ultimate reference in our relation to the world.

The thesis of this chapter, which intends to give language a naturalistic interpretation that allows us to speak of the "language" of genes, seems to be diametrically opposed to this position. According to the naturalistic interpretation, which is shared by other biologists, language is a natural principle for the organization of complex systems, which – in the words of Manfred Eigen (1979, p. 181) – "can be analyzed in an abstract sense, that is, without reference to human existence." From the standpoint of philosophical hermeneutics, such use of the word "language" is completely unacceptable. From this perspective, biologists who think and speak in this way about the existence of a "molecular language" look like drivers thundering down the motorway in the wrong direction – ignoring all the signposts naturally provided by human language for comprehending the world.

9.3 The "language" of genes

Impressive evidence for the naturalistic view of language seems to be found in the language-like arrangement of genetic information. Thus, as is well known, the genetic alphabet is grouped in higher-order informational units, which in genetic handwriting take over the functions of words, sentences, and so forth. And, like human language, genetic information has a hierarchical structure, which is unfolded in a complex feedback mechanism – a process that shows all the properties of a communication process between the genome and its physical context.

Of course, the parallels break down if we try to use the full riches of human language as a measure of the "language-like" structure of the genome. But from an evolutionary standpoint, there are good grounds to assert that "language" is indeed a natural phenomenon, which originates in the molecular language of the genome and has found, in the course of evolution, its hitherto highest expression in human language (Küppers, 1995). For evolutionary biologists, there is no question as to whether languages below the level of human language exist; the issue is rather about identifying the general conditions under which linguistic structures originate and evolve.

The significance of the natural phenomenon "language" for the explanation of living matter was recognized and first expressed with admirable clarity at the end of the nineteenth century by Friedrich Miescher, the discoverer of nucleic acids. Asking how a substance such as a nucleic acid can generate the vast diversity of genetic structures, he drew an analogy to the principles of stereochemistry. In the same way – Miescher argued – that a narrow variety of small molecular units is able to build up large molecules of almost unlimited complexity that are chemically very similar, but which have very different structures in three dimensions, the nucleic acids are capable of instructing the vast diversity of genetic structures. This line of thinking led Miescher to the conclusion that the nucleic acids must be able to "express all the riches and all the diversity of inheritance, just as words and ideas in all languages can be expressed in the 24–30 letters of the alphabet" (Miescher, 1897, p. 116). Obviously Miescher's view of living matter was that of a "linguistic movement" rather than that of a "clockwork machine." However, the "linguistic movement" of living matter is not a dead system of rules, but a dynamic one.

So, is this all just metaphoric speech? An outside observer, watching the disputants from a distance, might wonder what the controversy is all about, and might even suspect that it was a typical philosophers' war over the meaning of words. Our observer would be certain to draw attention to the fact that we repeatedly take words out of their original context and transpose them into another, so that any discourse about the world of nature is bound to employ metaphors, at least to a certain extent.

Why, then, should we not simply regard terms such as "information," "communication," and "language" in biology as what they really are: namely, adequate and highly resilient media for the description of the phenomena of life? Do the recent spectacular successes at the interface between biotechnology and information technology not justify the use of these concepts in biology? The construction of bio-computers, the development of genetic algorithms, the simulation of cognitive processes in neural networks, the coupling of nerve cells to computer chips, the generation of genetic information in evolution machines – all these would scarcely be conceivable without the information-theoretical foundations of living matter provided by biology.

However, the foregoing questions cannot be disposed of with simple arguments. This is above all because "information," "communication," and "language" are charged with other notions such as "meaning," "value," "truth," and the like. And this is where we run into the real nub of the discussion. Phenomena associated with meaning, as expressed in the semantic dimension of information, appear to evade completely all attempts to explain them on a naturalistic basis, and thus also to escape scientific description.

The right to interpret phenomena of meaning has traditionally been claimed by the humanities: especially by its hermeneutic disciplines. They have placed meaning, and thus also the understanding of meaning, at the center of their methodology; a clear demarcation against the natural sciences may indeed have been one of the motives for this. Whatever the reasons, the humanities have long gone their own way, have not considered it necessary to subject themselves to the scientific method of causal analysis – and have thus retained their independence for a considerable length of time.

The question of how broadly the concept of information may be applied is thus by no means a dispute about the content and the range of the applicability of a word. It would be truer to regard this question as the focal point at which philosophical controversies about the unity of knowledge converge – debates that have determined the relationship of the humanities and the natural sciences for more than a hundred years. The biological sciences, which stand at the junction between these two currents of thought, are always the first to get caught in the cross-fire. This is because an information-theoretical account of living matter involving a law-like explanation necessarily introduces questions of meaning and, thus, the semantic aspect of information (Küppers, 1996). Furthermore, the introduction of the semantic aspect of information in turn leads to the most fascinating plan-like and purpose-like aspects of living matter, which have every appearance of overstretching the capacity of traditional scientific explanation. Are, then, physical explanations – and with them the entire reductionistic research program in biology – doomed to founder on the semantic aspect of information?

9.4 The semantic dimension of information

Our discussion up to now has suggested that semantic information is "valued" information. The value of information is, however, not an absolute quantity; rather, it can only be judged by a receiver. Thus, the semantics of information depend fundamentally upon the state of the receiver. This state is determined by their prior knowledge, prejudices, expectations, and so forth. In short: the receiver's evaluation scale is the result of a particular, historically unique, pathway of experiences. Can – we may persist in asking – the particular and individual aspects of reality ever become the object of inquiry in a science based upon general laws and universal concepts? Even Aristotle addressed this important question. His answer was a clear "No." For him – the logician – there were no general discoveries to be made about things that were necessarily of an individual nature, because the logic of these two attributes – general and particular – made them mutually exclusive. This view has persisted through to our age, and has left a deep mark upon our present-day understanding of what science is and does.

Under these circumstances, the achievement of the philosopher Ernst Cassirer appears all the more admirable. Opposing the Aristotelian tradition, Cassirer attempted to bridge the presumed gap between the general and the particular (Cassirer, 1910). Particular phenomena, he argued, do not become particular because they evade the general rules, but because they stand in a particular – that is, singular – relationship to them. Cassirer's reflections may have been triggered by an *aperçu* of von Goethe (1981, p. 433): "The general and the particular coincide – the particular is the general as it appears under various conditions."

According to Cassirer, it is the unique constellation of general aspects of a phenomenon that makes up its uniqueness. This is an interesting idea. It makes clear that an all-embracing theory of semantic information is impossible, whereas general aspects of semantics can very well be discerned. Taken for themselves, these aspects may never completely embrace the phenomenon in question. But through their unique interconnectedness, they allow the particular characteristics of the phenomenon to show through clearly. In other words: the unique properties of semantic information originate by superposition of its general disposition. The aspects that constitute semantic information in this sense include, among others, their novelty and their pragmatic relevance as well as their complexity (Küppers, 1996).

At the beginning of the 1950s, the philosophers and logicians Yehoshua Bar-Hillel and Rudolf Carnap (1953) tried to quantify the meaning of a linguistic expression in terms of its novelty value. This idea was a direct continuation of the concept developed within the framework of Shannon's information theory, where the information content of a message is coupled to its expectation value: the lower the expectation value of a message, the higher its novelty and thus its information content. This approach takes care of the fact that an important task of information is to eliminate or counteract uncertainty. However, the examples adduced by Bar-Hillel and Carnap are restricted to an artificial language.

A more powerful approach to measuring the semantics of information is that based upon its pragmatic relevance. This approach has been described in a paradigmatic way by Donald MacKay (1969) in his book *Information, Mechanism and Meaning*. The pragmatic aspect of information refers to the action(s) of the receiver to which the information leads, or in which it results.

For some time now, my own efforts have been focused on a new approach, intended to investigate the complexity of semantic information (Küppers, 1996). Unlike the approaches described above, this one does not seek to make the meaning of information directly amenable to measurement. Rather, it aims to demarcate the most general conditions that make up the essence of semantic information. Investigations of this kind are important because they afford a more general insight into the question of the origin of information, and therefore have consequences for major fundamental problems of biology such as the origin and evolution of life (Küppers, 2000a).

9.5 How does information originate?

Let us consider the relationship between semantic information and complexity in more detail. Information, as we have said, is always related to an entity that receives and evaluates the information. This in turn means that evaluation presupposes some other information that underlies the process of registration and processing of the incoming information. But how much information is needed in order to understand, in the foregoing sense, an item of incoming information? This question expresses the quantitative version of the hermeneutic thesis, according to which a person can only understand some piece of information when it has already understood some other information.

At first sight, it would seem impossible to provide any kind of answer to this question since it involves the concept of understanding, which, as we have seen, is already difficult to understand by itself, let alone to quantify. Surprisingly, however, an answer can be given, at least if we restrict ourselves to the minimal conditions for understanding. To this belongs first of all the sheer registration by the receiver of the information to be understood. If the information concerned conveys meaning – that is, information of maximum complexity – then the receiver must obviously record its entire symbol sequence before the process of understanding can begin. Thus, even the act of recording involves information of the same degree of (algorithmic) complexity as that of the symbol sequence that is to be understood.

This surprising result is related to the fact that information conveying meaning cannot be compressed without change in, or even loss of, its meaning. It is true that the contents of a message can be shortened into a telegram style or a tabloid headline; however, this always entails some loss of information. This is the case for any meaningful information: be it a great epic poem or simply the day's weather report. Viewed technically, this means that no algorithms – that is, computer programs – exist that can extrapolate arbitrarily chosen parts of the message and thus generate the rest of the message. But if there are no meaning-generating algorithms, then no information can arise de novo. Therefore, to understand a piece of information of a certain complexity, one always requires background information that is at least of the same complexity. This is the sought-after answer to the question of how much information is needed to understand some other information. Ultimately, it implies that there are no "informational perpetual-motion machines" that can generate meaningful information out of nothing (Küppers, 1996).

This result is the consequence of a rigorous relativization of the concept of information. It is a continuation of the development that characterized the progress of physics in the last century: the path from the absolute to the relative. This began with the abandoning of basic concepts that had been understood in an absolute sense – ideas such as "space," "time," and "object" – and has since led to well-known and far-reaching consequences for the foundations of physics. Whether the thorough-going relativization of the concept of information will one day lead to a comparable revolution in biological thinking cannot at present be said. This is largely due to the fact that the results up to now have been derived with respect to the semantic dimension of human language, and it is not yet clear to what extent they are applicable to the "language of genes." For this reason, questions such as whether evolution is a sort of perpetual-motion machine must for the present remain open.

At least it is certain that we must take leave of the idea of being able, one day, to construct intelligent machines that spontaneously generate meaningful information de novo and continually raise its complexity. If information always refers to other information, can then information in a genuine sense ever be generated? Or are the processes by which it arises in nature or in society nothing more than processes of transformation: that is, translation and re-evaluation of information, admittedly in an information space of gigantic dimensions, so that the result always seems to be new and unique? Questions such as these take us to the frontline of fundamental research, where question after question arises, and where we have a wealth of opportunities for speculation but no real answers.

9.6 The world of abstract structures

Finally, I should like to return briefly to the question with which we began: Are the ideas of "information," "communication," and "language" applicable to the world of material structures? We saw how difficult it is to decide this on a philosophical basis. But it may also be the case that the question is wrongly put. There does indeed seem a surprising solution on the way: one prompted by current scientific developments. In the last few decades, at the border between the natural sciences and the humanities, a new scientific domain is emerging that has been termed "structural sciences" (Küppers, 2000b). Alongside information theory, it encompasses important disciplines such as cybernetics, game theory, system theory, complexity theory, network theory, synergetics, and semiotics, to mention but a few. The object of structural sciences is the way in which the reality is structured – expressed, investigated, and described in an abstract form. This is done irrespectively of whether these structures occur in a natural or an artificial, a living or a non-living, system. Among these, "information," "communication," and "language" can be treated within structural sciences as abstract structures, without the question of their actual nature being raised. By considering reality only in terms of its abstract structures, without making any distinction between objects of "nature" and "culture," the structural sciences build a bridge between the natural sciences and the humanities and thus have major significance for the unity of science (Küppers, 2000b).

In philosophy, the structural view of the world is not new. Within the frame of French structuralism, Gilles Deleuze took the linguistic metaphor to its limit when he said that "There are no structures that are not linguistic ones . . . and objects themselves only have structure in that they conduct a silent discourse, which is the language of signs" (Deleuze, 2002, p. 239). Seen from this perspective, Gadamer's dictum "Being that can be understood is language" (Gadamer, 1965, p. 450) takes on a radically new meaning: "Being" can only be understood when it already has a linguistic structure. Pursuing this corollary, the philosopher Hans Blumenberg (2000), in a broad review of modern cultural history, has shown that – and how – the linguistic metaphor has made possible the "readability" (that is, the understanding) of the world. However, the relativity of all understanding has of necessity meant that the material "read" was reinterpreted over and over again, and that the course of time has led to an ever more accurate appreciation of which "readings" are wrong. In this way, we have approached, step by step, an increasingly discriminating understanding of the reality surrounding us.

References

Bar-Hillel, Y., and Carnap, R. (1953). Semantic information. *British Journal for the Philosophy of Science*, 4: 147.

Blumenberg, H. (2000). *Die Lesbarkeit der Welt*. Frankfurt/Main: Suhrkamp.

Cassirer, E. (1910). *Substanzbegriff und Funktionsbegriff*. Berlin: Bruno Cassirer.

Deleuze, G. (2002). À quoi reconnaît-on le structuralisme? In *L'Île Déserte et Autres Textes*, ed. D. Lapoujade. Paris: Minuit, 238–269.

Eigen, M. (1979). Sprache und Lernen auf molekularer Ebene. In *Der Mensch und seine Sprache*, eds A. Peisl and A. Mohler. Frankfurt/Main: Propyläen Verlag, 181–218.

Gadamer, H.-G. (1965). *Wahrheit und Methode*, 2nd ed. Tübingen: J. B. C. Mohr.

Küppers, B.-O. (1990). *Information and the Origin of Life*. Cambridge, MA: The MIT Press.

Küppers, B.-O. (1992). Understanding complexity. In *Emergence or Reduction?* eds A. Beckermann, H. Flohr and J. Kim. Berlin: de Gruyter, 241–256 [reprinted in *Chaos and Complexity*, eds R. J. Russell, N. Murphy and A. R. Peacocke (1995). Vatican City State: Vatican Observatory Publications, 93–105].

Küppers, B.-O. (1995). The context-dependence of biological information. In *Information. New Questions to a Multidisciplinary Concept*, eds K. Kornwachs and K. Jacoby. Berlin: Akademie Verlag, 135–145.

Küppers, B.-O. (1996). Der semantische Aspekt von Information und seine evolutionsbiologische Bedeutung. *Nova Acta Leopoldina*, 294: 195–219.

Küppers, B.-O. (2000a). The world of biological complexity: Origin and evolution of life. In *Many Worlds*, ed. S. J. Dick. Pennsylvania: Templeton Foundation Press, 31–43.

Küppers, B.-O. (2000b). Die Strukturwissenschaften als Bindeglied zwischen Natur- und Geisteswissenschaften. In *Die Einheit der Wirklichkeit*, ed. B.-O. Küppers. Munich: Fink Verlag, 89–105.

MacKay, D. M. (1969). *Information, Mechanism and Meaning*. Cambridge, MA: The MIT Press.

Miescher, F. (1897). *Die histochemischen und physiologischen Arbeiten*. Bd. I. Leipzig: F. C. W. Vogel.

von Goethe, J. W. (1981). *Werke*, Hamburger edition vol. 12. München: C. H. Beck.

Semiotic freedom: an emerging force

JESPER HOFFMEYER

~

The term "information" has become nearly omnipresent in modern biology (and medicine). One would probably not exaggerate if the famous saying of evolutionary biologist Theodosius Dobhzhansky, that "nothing in biology makes sense except in the light of evolution" should nowadays be reframed as "nothing in biology makes sense except in the light of information." But are those two concepts, evolution and information, somehow internally related? And if so, how?

10.1 Information in evolution

In textbooks, newspapers, and even scientific papers, the meanings of the terms "evolution" and "information" are generally supposed to be well known, and they are rarely explained. And yet, there is no general consensus in science, or even in biology, about what they really mean. For instance, the preferred exemplar (in the Kuhnian sense) of evolution by natural selection is that of industrial melanism. In woodlands, where industrial pollution has killed the lichens and exposed the dark brown tree trunks,

Information and the Nature of Reality: From Physics to Metaphysics, eds. Paul Davies and Niels Henrik Gregersen. Published by Cambridge University Press © P. Davies and N. Gregersen 2010, 2014.

dark forms of the peppered moth – melanics – are supposedly better camouflaged against predation from birds
than are the light gray forms that predominated before
the Industrial Revolution. The observation by 1950 that
darker forms had largely displaced lighter forms was thus
taken as evidence for natural selection in action. This
exemplar does indeed illustrate the effect of natural selection, but whether it shows evolution depends on your idea
of evolution. Thus, if by the term "evolution" is meant
something like "the origin of species," then it is remarkable that no speciation has actually taken place here, and
it is not obvious that any speciation would indeed occur
in a case like this. So, populations do indeed change as
a result of natural selection, but is this mechanism also
behind speciation and evolution at large? The majority of
biologists certainly feel assured that macroevolution – that
is, evolution above the level of species – is in fact the tardy
result of an infinitely ongoing microevolution (adaptation
in populations), but there remains serious disagreement
on this (Depew and Weber, 1995; Gould, 2002).

Likewise, the meaning of the term "information" is
loaded with ambiguities. Whereas theorists may eradi-
cate such ambiguities by rigorous definition, it is often
less than clear how well-defined concepts of information
relate to the actual use of information as an explanatory
tool in biology. When talking about information, most
biologists probably have in mind the kind of information
expressed in the so-called "central dogma," as stated by
Francis Crick in the following terms: "Once information
has passed into protein it cannot get out again"; that is, the
flow of information in the cell is unidirectional; it originates in the gene and ends up as protein (see Figure 10.1).

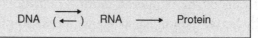

FIG. 10.1 According to the "central dogma," the information flow in a cell is one-way. Information is passed from DNA to RNA and further to protein, but never in the reverse direction.

Writers of textbooks are anxious to point out that the kind of information we are talking about here is "sequential information," which is "replicated" when the cell divides, is "transcribed" into mRNA, and is "expressed" by genes. Following transcription to mRNA, the information is said to be "processed" or "edited," whereupon it "migrates" from the nucleus to the cytoplasm, where it is finally "read" by a ribosome that "translates" it into protein. However, as was shown by Sahorta Sarkar, nothing really goes on here that could not be expressed by decent biochemical processes exhibiting traditional efficient causality (Sarkar, 1996). So, why do molecular biologists prefer to talk about "information"?

There can, certainly, be no doubt about the heuristic value of the information metaphor; something is added to our understanding when we talk about information rather than just about chemistry. In fact, without the information metaphor, it would be hard to understand modern biochemistry at all.[1] But whatever this added

[1] One promising candidate for an information concept that is both rigorous and biologically useful has been developed in ecology as the so-called "average mutual information," a measure for how well organized or determinate a configuration of ecological relationships appears. Theoretical ecologist Robert Ulanowicz has developed this concept further to a concept of *ascendency* that represents: "the coherent power a system could bring to bear in ordering itself and the world around it" (Ulanowicz, 2009).

understanding consists of, it is not part of the cellular reality according to molecular biologists. At the cellular level chemistry exhausts what goes on. The reason for this insistence on the reducibility of information to chemistry, I shall suggest, is that the heuristic value of the information concept is connected to the role that history (evolution) plays in the life of cells and organisms. What happens is that "history talks", but history is not considered part of biochemistry or molecular biology. As shown by Terrence Deacon, in Chapter 8 of this volume, the contextual (historical) aspect of information is due ultimately to the fact that informative signals are necessarily caused by externally derived perturbations of some medium away from its expected state. The receipt of an informative signal ipso facto provides evidence of the material influence of something other than the signal itself, linking its resultant form to this "absent cause": the immediate or mediating object of reference. Although history is, of course, not "doing" anything, and most certainly does not "talk", the present form of the informative substrate – for example, the actual sequence of nucleotides in a section of the DNA string – does nevertheless refer to absent causes: that is, causes that are connected to the evolutionary past of the species. This referential aspect of information furthermore provides for the inherent intentionality of biological information to the extent that it reflects the workings of natural selection. As Deacon would put it, genetic and phenotypic representations with the least correlation with environmental regularities will not be transmitted (reproduced).

The (subconscious?) effect of the apparently unavoidable "information talk" in molecular biology may thus be to smuggle in the intentional (semiotic) aspect of information by the back door, so to say. At the surface, genetic information is treated as if it was just a simple causal factor, but its deep appeal to our understanding derives from its hidden connotation of an otherwise tabooed intentionality. The exclusion of history may be relatively innocent at the level of molecular biology itself, but when molecular biology is understood as the basis for genetics and evolution, this ahistorical understanding serves to reify processes that are in fact embedded in contextual constraints that should be accounted for in our theories. Genes are supposed to specify particular traits or characters such as missing eyes in a fruit fly or Huntington's chorea in human beings. So the question is, what does it mean that genes are carrying information about such properties?

The identification of genetic information with "sequential information," however, reinforces and is reinforced by the belief in genocentric versions of Darwinism. If information is nothing but a molecular property of DNA, it may be replicated and transported down through generations in well-defined and unambiguous units open to modification by the combined processes of mutation and selection. Natural selection will then gradually optimize the set of instructional units carried in the gene pool of a population. What is pushed aside in this simplified view of gene function once again is the burden of context. The concrete functioning of genes is dependent not only on the genetic background on which they happen to be placed, but also on a host of circumstances normally described collectively as the milieu (Griffiths and Gray, 1994; Gray, Griffiths, and Oyama, 2001). The conception of genes as unambiguous or autonomous functional units does not even come true in those monogenetic diseases that originally served as models for our ideas of gene function. Thus, the monogenetic disease PKU has now been shown to exhibit quite unexpected phenotypic variation (Scriver and Waters, 1999). Not all untreated carriers of the "disease gene" exhibit disturbed cognitive development, probably because the build-up of toxic concentration of the amino acid phenylalanine in the brain is influenced by unknown factors. Genetic information does not simply "cause" things to happen.

Needless to say, proponents of the "intelligent design" movement use this kind of criticism to attack the legitimacy of evolutionary theory. And Darwinists, on their side, respond with deep suspicion towards any criticism of the general scheme of evolutionary theory. A third possibility exists though, and this is the possibility I am going to present and discuss in the remaining part of this chapter: namely, the approach called *biosemiotics*.

10.2 Semiosis and life

Biosemiotics suggests that living systems should be studied as semiotic[2] systems in their own right. This idea is based on the belief that the poverty of the information discourse in biological sciences results from the reductive neglect of the interpretative aspect of biological information. By introducing the concept of the sign, as developed by US chemist and philosopher Charles Sanders Peirce (1839–1914) as a substitute for information, it will be assured that the interpretative side of information is not

[2] Semiosis = sign activity. Semiotics = the sciences studying sign activity.

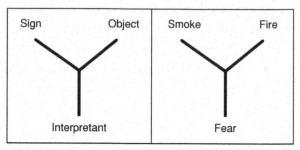

FIG. 10.2 Sign action. Left: a graphical illustration of Charles Sanders Peirce's conception of a sign as a triadic relation connecting the primary sign to its object through the production of an interpretant. Right: Smoke provokes the formation of an interpretant in the brain of an animal, causing it to flee. The animal is seized by alarm, and this being "seized by alarm" is the interpretant. In this figure, the three instances of the sign relation are connected by a tripod rather than by a triangle in order to emphasize the internal logic of the sign relation, which should never be confused with a mere summation of three relations between corners in a triangle.

neglected. In everyday parlance, a sign is simply "something that refers to something else", like smoke refers to a fire. This reference, however, cannot be brought about without a process of interpretation. Thus, a baby will not know that smoke signifies fire. A deer, on the other hand, will perhaps not know that smoke refers to fire, but it definitely takes smoke as a sign of danger. And adult people, of course, usually turn their head in order to see where the fire might be.[3] An "interpretant" is constructed by brain processes that mediate the connection between the sense impression of smoke and the presumed existence of something burning. Thus, we get the Peircean triadic sign concept, as given in Figure 10.2.

---

[3] But indeed, inside a theater the smoke might just be faked. Semiotics, as Umberto Eco once said, is the science of lying.

Signs, however, are not causes in the traditional sense of (Aristotelian) efficient causality, for the effect of a sign is not compulsory but depends upon a process of interpretation, and the interpretation may well be – and probably most often is – "mistaken." An example of this is when a predator is lured away from a bird's nest because it misinterprets the clumsy behavior of the bird as a sign for an easy catch. The bird, however, just pretended to have a broken wing and flew away the minute the predator had missed the nest. In this semiotically interesting case, the bird profits from its "knowledge" of the predator's predictable habit of going for the easy catch. Whether the bird's "knowledge" in this case is phylogenetically (instinctually) or ontogenetically (learned) rooted does not change the general logic of the situation – what differs is only a question of the implied mechanism and the time span of the interpretative act. On the one hand, if the bird's behavior is instinctually based, it is the result of an evolutionary interpretation, and we are then talking of time spans of perhaps millions of years during which this particular habit developed in the bird as an interpretative response to the predictable behavior of predators. If, on the other hand, the behavior is based on learning, the interpretative act is a product of brain processes not far from the brain processes involved in our own human interpretative activity.

I said above that the heuristic value of the information concept derives from the historical nature of living systems. Although nobody in science denies that living creatures are the result of an evolutionary process, this process is normally not conceived as historical in the usual sense of this term, because it is assumed to have obeyed the deterministic rule of natural laws. The metaphysics of determinism, however, seem less and less supported by the findings of modern science. It cannot, of course, be excluded right away, but there no longer seems to be a strong reason to adopt it. In Stuart Kauffman's book *Investigations* (2000), an important part of the analysis turns on the question of the non-ergodicity of the universe, meaning that the universe never had the time it would have needed should its present state of affairs in any way be representative of its in-built possibilities. The persistent movement of the universe into the "adjacent possible" precludes it from ever reaching a state that depends on statistical likelihood. Instead, the universe is historical, for "History enters when the space of the possible that might have been explored is larger, or vastly larger, than what has actually occurred" (Kauffman, 2000, p. 152).

The historical nature of the world has profound consequences for the study of life, because it confronts us with the problem of organization in a new way. If the complex forms of organization exhibited by living systems – from the cell to the ecosystem – are not the inescapable result of predictable lawfulness, they must instead have emerged through processes that are still in need of discovery. The principle of natural selection, of course, greatly helps us in explaining the widespread adaptedness of biosystems, but we need an additional principle that would solve the fundamental question of the "aboutness" of life, the never-ending chain of attempts by living systems to come to terms with their conditions of life. As Stuart Kauffman and Philip Clayton have formulated this: "it is a stunning fact that the universe has given rise to entities that do, daily, modify the universe to their own ends. We shall call this capacity *agency*" (Clayton and Kauffman, 2006, p. 504).

This "agency" necessarily involves a kind of measuring process, whereby the agent is enabled to modify its environment in a selective way, furthering its own ends, and this measuring process is the core of the striking "aboutness" that characterizes living systems. "Aboutness," thus, is not derivable from the principle of natural selection for the simple reason that it is required for natural selection to operate in the first place.[4] If organisms did not exhibit aboutness, if they did not "take an interest" in the world around them (if they did not "strive" – to use Darwin's own term), there would be no "competition for survival" but only disorganized activity leading nowhere.

However, if natural selection is not responsible for this aboutness, what then is the basis of it? Kauffman and Clayton propose a tentative five-part definition:

a minimal molecular autonomous agent: such a system should be able to reproduce with heritable variation, should perform at least one work cycle, should have boundaries such that it can be individuated naturally, should engage in self-propagating work and constraint construction, and should be able to choose between at least two alternatives

(Clayton and Kauffman, 2006)

Terrence Deacon has suggested an even more simple system as a candidate for the prebiotic emergence of life and agency, a system that he calls the "autocell" (Deacon and Sherman, 2008). Autocells are self-assembling molecular structures that derive their individuality from a synergistic relationship between two kinds of self-organizing

---

[4] In the simplest case, the intentional dynamics of aboutness shows itself as fertility.

processes that reciprocally depend upon one another's persistence. Such autocells could have been an important stepping stone in the process leading from the non-living to the living. However, as Deacon and Sherman themselves point out, autocells are not yet full-blown living systems. They lack several features that are generally considered criteria for being alive, such as the possession of the replicative molecules of RNA or DNA, and differential survival through replications. Furthermore, autocells will not meet the set of criteria put forward by Kauffman and Clayton. The autocell model, however, does demonstrate a possible unbroken continuity from thermodynamics to evolvability.

The difficult problem to solve in any theory of the origin of agency and life is how to unify two normally quite separate kinds of dynamics: a dynamics of chemical interaction patterns and a dynamics of signification or semiosis. This immediately places this question in the contextual situation of the environment. The Kauffman–Clayton criteria clearly do so, but they also presuppose a far more complex beginning than Deacon's autocell model. It remains to be seen if, or how, these two approaches might be reconciled.

Biosemiotics, of course, immediately reminds us that "to ask for the origin of life is to ask for the origin of the environment" (Hoffmeyer, 1998). Living organisms are inscribed in their environments much like patterns woven into a carpet – the two cannot get apart. From a semiotic point of view, the decisive step in the process that led to the origin of life was the appearance in the world of a new kind of asymmetry, *an asymmetry between inside and outside*. The formation of a closed membrane around an autocatalytically closed system of components (Kauffman, 1993) might have been an initial step. Such a membrane would have created what is probably the most essential and unique characteristic of life: the never-ending *interest* of the insides into their outsides or, in other words, cellular *aboutness*. I have suggested that this "interest" should be understood as a property that ultimately was derived from the primordial membrane itself. A closed membrane sheet necessarily has two distinct kinds of exterior: the inside exterior and the outside exterior. For agency to appear, such membranes must have managed not only to canalize a selective flow of chemicals across themselves, but also to subsume their interior system of components to help them resist the flow of perturbations from the outside exterior. Or, in other words, for the membranes to persist, they would have to function as interfaces connecting their inside worlds to the outside world. At some point in pre-biotic development, a self-referential digital description (in RNA or DNA) of constitutive "cell" components was established. For a prebiotic system to become a true living system, however, this self-referential description of the system had to be integrated (for example through signal transduction) into the other-referential system of receptors at the surface of the cell. *Such a stable integration of a self-referential digitally coded system into an other-referential analogically coded system may perhaps be seen as a definition of life* (Hoffmeyer, 1998).

10.3 Semiotic freedom

When a bacterial cell finds itself in a gradient of nutrients and swims right instead of left, the cell is making a choice.[5] The choice is of course based on a complicated chemotactical machinery (comprising some scores of different protein species), but a biochemical analysis of the chemotactic system does not exhaust our need for understanding. We also need to know why this apparatus was developed in the first place. Biochemistry essentially helps us to construct an image of cellular life as chains or webs of chemical reactions taking place inside a cell or an organism. This of course is extremely useful in many contexts, but in itself, it does not contain information about the structural logic that has ruled the organization of this apparent mess of millions of chemical reactions taking place in a cell. As Nobel laureate and biochemist Alfred Gilman once told *Scientific American*:

I could draw you a map of all the tens of thousands of components in a single-celled organism and put all the proper arrows connecting them (and even then) I or anybody else would look at that map and have absolutely no ability whatsoever to predict anything.   (Gibbs, 2001, p. 53)

---

[5] Again I am here using teleological language. The idea is not, of course, that the bacterium makes a conscious choice, but only that it systematically favors a distinct response out of several equally possible behaviors. If it is objected that this "preference" is not a real "preference" or "choice" because the bacterium could not possibly have failed to respond the way it did, I will contend that: (1) considered as an evolving species the bacterium does in fact have a choice, in the sense that it might have evolved differently; and (2) even single bacterial cells are complex systems that exhibit truly unpredictable (chaotic) behavior and, in fact, mutant cells might behave differently in the same situation. It should be noticed here that the idea of an original "wild-type" bacterium has now been given up: all bacterial cells are thus "mutants" or "normal," depending on how the context is defined.

The key to cellular or organismic organization must be searched for in the historical (evolutionary) constraints on the interaction of simple biosystems with each other and with their environments. In the semiotic understanding outlined above, the chemotactic machinery serves to integrate the sensing of the outer world to the reality of the inner world as this reality is described in the self-referential, or genetic, system. Natural selection, of course, has modulated this system all the way down from the first cells on Earth; and for all we know, the system is very well safeguarded. The possibility remains, however, that mutations will spoil or change it and, more importantly perhaps, that external factors may fool it. Researchers may, for instance, easily fool the bacterial chemotactic system by adding nutrient analogs (such as artificial sweeteners instead of glucose) to the medium; and most likely nature itself will from time to time "invent" comparable kinds of dupe.

In such cases, it seems appropriate to say that the cell *misinterprets* the chemical signs of its environment. Such misinterpretations are dangerous, and natural selection will favor any solution that helps the organism to better interpret the situations it meets. Indeed, selection would be expected to favor the evolution of more sophisticated forms of "semiotic freedom" in the sense of an increased capacity for responding to a variety of signs through the formation of (locally) "meaningful" interpretants. Semiotic freedom (or interpretance) allows a system to "read" many sorts of "cue" in the surroundings, and this would normally have beneficial effects on fitness. Thus, from the modest beginnings we saw in chemotactic bacteria the semiotic freedom of organic systems would have tended to increase, and although it has not been easy to prove that any systematic increase in complexity, as this concept has traditionally been defined, has in fact accompanied the evolutionary process, it is quite obvious that semiotic complexity or freedom has indeed attained higher levels in later stages, advanced species of birds and mammals in general being semiotically much more sophisticated than less advanced species (Hoffmeyer, 1996).

Allowing for semiotic freedom in the organic world significantly changes the task of explaining emergent evolution, because semiotic freedom has a self-amplifying dynamic. Communicative patterns in assemblies of cells or individuals may often have first appeared as a simple result of the trial-and-error process of normal interaction, and may then endure for considerable periods of time. If such patterns are advantageous to the populations (cells or organisms), they may eventually become scaffolded by later mutational events. Through this "semi-Baldwinian" mechanism, the evolutionary process will enter a formerly forbidden area of goal-directedness (Hoffmeyer and Kull, 2003).

Biosemiotics presents a strong argument for an emergentist view of life. By semiotic emergence, I mean the establishment of macro-entities or higher-level patterns through a situated exchange of signs between sub-components. The important point here is that whereas the emergence of higher-level patterns may seem to be slightly mysterious (often raising suspicion of vitalism), as long as only physical interactions between entities are considered, the same outcome becomes quite understandable when based on *semiotic* interactions among entities at the lower level. Most importantly, *semiotic emergence* in this sense may stand as a possible alternative candidate to natural selection as a mechanism for explaining the evolution of purposive behavior.

The biosemiotic understanding also implies that semiosis cannot be used to mark off the human species from the rest of the world's creatures. Our species' linguistic skills make us very different from other species on Earth, indeed, but as shown in the work of Terrence Deacon, the capacity for semiotic reference is not in itself the distinctive mark between humans and animals. What is distinctive is our unmatched talent for that particular kind of semiotic activity that is symbolic reference (Deacon, 1997), and thus languaging. We share the semiotic capacity as such (that is, the iconic and indexical referencing) with all life forms, and biosemiotics thus puts us back into nature in the same time as it reconstructs nature as a place for humans to belong. Nature is much more like us than science – in its obedience to the anthropocentric taboo – has allowed. Instead of the Cartesian either-or thinking, biosemiotics institutes a more-or-less thinking.

John Deely has called the human being "the semiotic animal" (Deely, 2007). Semiosis – sign action – takes place all over the life sphere, but only humans *know* the difference between signs and things; only humans are semiotic animals.

10.4 Semiotic emergence and downward causation

A key question in discussions of emergence concerns the ontological reality of causative influences upon lower-level entities exerted by the macrostate of the system. The expression "downward causation" is often used in this type of causal relationship in which a macrostate acts upon the very microstates of which it consists (Campbell, 1974). The term may be seen as an attempt to express parts of what used to be called "final causation" without linking ourselves to the cultural inheritance of Aristotelianism and all the baggage this implies. As we shall see, semiotic emergence and downward causation are two sides – or rather two aspects – of the same coin, which I would prefer to call "semiotic causation," bringing about effects through a process of interpretation.

Deborah Gordon's laborious and highly rewarding work with ants of the species *Pogonomyrmex barbatus* (who live in a harsh zone bordering the deserts between Arizona and New Mexico) is illustrative here. Gordon's work revealed that the survival of colonies of this species is deeply dependent on the regulation of a sophisticated pattern of semiotic interactions between individual ants, which then raises the question of whether the ant colony deserves to be seen as a "superorganism." (Gordon, 1995, 1999). Seen from a semiotic point of view, a superorganism might be understood as an assembly of organisms that collectively interacts with its environments in a way that depends on a finely elaborated internal semiotic activity among the individual organisms – a "proto-endosemiotics" (Hoffmeyer, 2008). Gordon found that a particularly important element in the colony's behavior and growth process is what she calls "job allocation" – and she shows that although this task does indeed rely on a quite schematic interaction pattern between different groups of ants, an element of unpredictability persists:

An ant does not respond the same way every time to the same stimulus; nor do colonies. Some events influence the probabilities that certain ants will perform certain tasks, and this regularity leads to predictable tendencies rather than perfectly deterministic outcomes.   (Gordon, 1999, p. 139)

Gordon's experiments in this area may be seen as a response to experiments performed by the founder of sociobiology, Edward O. Wilson, that were claimed to show a full-blown determinism in the response pattern of ants to chemical signals (for example oleic acid (Wilson, 1975)). Gordon's experiments, on the contrary, showed that "Just as the same word can have different meanings in different situations...so the same chemical cue can elicit different responses in different social situations" (Gordon, 1999, p. 97). Physiological, social, and ecological processes are simultaneously at work, says Gordon, and none of them is more basic than the others: "Living levels of organization is central to any study of social behavior. For humans and other social animals, an individual's behavior is always embedded in a social world" (ibid., p. 96).

The semiotic competence of subunits, then – whether these subunits are human individuals in a society, plants in an ecosystem, cells in a multicellular organism, or ants in an ant colony – is the medium through which the behavior and integrity of the higher-level entity is produced and maintained. To the extent that such a system's endosemiotic relations perceive and utilize cues and signs that indicate (are indices for) the state of the holistic unit and its "needs," it seems justified to talk about these processes as genuinely *endosemiotic* – and

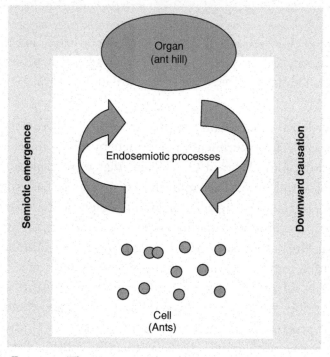

FIG. 10.3 The connection between semiotic emergence and "downward causation." Downward causation operates through indexical sign relations; that is, the values of system parameters are interpreted by lower-level agents as indexical signs. But this state of affairs in itself presupposes the formation in the first place of a large-scale pattern with a behavior that stabilizes the semiotic interaction between parts (from Hoffmeyer, 2008).

consequently, the holistic system itself deserves to be ascribed a status as an autonomous unit: a superorganism.

The evolutionary formation of this kind of autonomous macro-entity is the quintessence of what is called "downward causation" as defined above, and it is suggested that the connection between semiotic emergence and downward causation is taken as constitutive for both phenomena, as shown in Figure 10.3.

The semiotic relations between subunits that collectively account for the stability of the large-scale or holistic system (for instance the ant hill, the multicellular organism, or perhaps the symbiotic system of bobtail squids and light-emitting *Vibrio* bacteria: see below) must necessarily be geared to respond to changes in the environment in ways that do not threaten the integrity of the large-scale system. Subunits, for their part, must receive messages telling them how to uphold the macro-system, and the easiest way to do this is probably to distribute the needs of the macro-system via "indexical" signs.

An interesting example is the so-called "quorum sensing." Quorum sensing has become the designation for a kind of communicative activity in bacteria in which the density of bacteria present is a causal factor. In short, quorum sensing is the result of a process in which each single bacterium excretes a certain chemical compound such that the concentration of this compound in the medium will reflect the number of bacteria per unit volume. Quorum sensing occurs if the compound, after having reached a threshold concentration, binds to a regulatory protein in the cell and thereby initiates the transcription of specific genes. An illustrative case of quorum sensing occurs in a species of squid, *Euprymna scolopes*, which hunts small fish by night on the coral reefs off the coast of Hawaii.[6] Moonlight causes the squid to cast a shadow that makes them an easy catch for predators. As a defense strategy, *E. scolopes* has evolved a sophisticated way of emitting light that effectively "hides" its own shadow. "Counter-illumination" is

---

[6] For a detailed account of the semiotics of this case, see Bruni (2007) and Hoffmeyer (2008).

the name given to this kind of camouflage, and it is only made possible by the squid's symbiotic relationship with luminous bacteria called *Vibrio fischeri* that live in the mantel cavity of the squid. Living on food provided by the digestive system of the squid, the bacteria emit light of the exact same intensity and color as the light reaching the squid from the Moon, and this prevents predators from seeing the squid from below (McFall-Ngai and Ruby, 1998).

In the morning, the squid bury themselves in the sand and excrete 90–95% of the bacteria, which brings the density of the remaining bacteria well below the threshold level. The bacteria continue producing the particular substance, *N*-acyl homoserine lactone, which triggers the induction of light emission, but because the density of bacteria is low the concentration of this substance remains too low to elicit any induction of the involved lux genes. Apparently, the squid is in full control of the bacterial growth rate by adjusting the supply of oxygen, and at sunset the population of bacteria reaches the threshold level once again. When night approaches, the squid makes sure that the bacterial density in its mantel cavity (and thus the *N*-acyl homoserine lactone concentration) is high enough for the bacteria to respond by allowing for the transcription of lux operon genes – and thus for light emission – to start. The point I want to make here is that the semiotic emergence whereby this system was gradually established in the course of evolution necessarily also demanded the invention of semiotic means for assuring the stability of the system, and these "semiotic means" are precisely what we understand by "downward causation." And here I have only touched upon the surface of the complexities of the communicative network operative in bringing this symbiotic interaction between the squid and the bacterium to work.

Downward causation and semiotic emergence are thus two interwoven, but not identical, aspects of the same process. That this mechanism, based on indexical semiosis, is indeed coupled to the circadian rhythms of the squid was confirmed by the finding that the squid has means at its disposal to fine-tune light emission. It may, for instance, change the wavelength of the emitted light by help of a "yellow filter," and it may weaken the intensity of light by opening a bag of "ink" (Bruni, 2007). Presumably, the indexical sign process itself is too slow in its effects to compensate for momentary variations in background light (as occurs under cloudy conditions), and evolution therefore had to provide the macro-system (the squid) with a number of "additional screws" for fine-tuning.

10.5 Biosemiotics and God

Contrary to the genocentric neo-Darwinian orthodoxy, the view of animate nature as profuse with signs does not contradict the fact that our Earth has supported the evolution of conscious human beings with moral feelings. Biosemiotics may potentially account for the appearance of such creatures through immanent natural processes. The biosemiotic approach thereby overturns the need for, or legitimacy of, the argument for intelligent design. Biosemiotics does not logically entail any stance on the presence or absence of a transcendental creator, but it may be seen as rescuing the scientific world view from the need for admitting such a transcendental power, a need that seems rather inescapable if one adheres to a traditional scientific perspective. As Deacon and Sherman have put it:

In the debate between science and fundamentalist religion that the world's citizens watch so attentively, the burden is on science to demonstrate how purposive processes can emerge in the absence of antecedent intelligence, carefully selected prior conditions, or intrinsically teleological components.

(Deacon and Sherman, 2008, pp. 59–76)

Deacon and Sherman have made an impressive attempt at "tracking the emergence of the fledgling precursor to these intentional relations," but although their approach is perfectly naturalistic, its emphasis on theories of complexity and semiotics unfortunately brings it well outside the narrow space of accepted causative agencies in mainstream scientific world view.

The present author remains agnostic towards the questions of a transcendent or immanent deity, but notes that such agnosticism would have been hard or impossible to maintain, had it not been for the biosemiotic solution to the hard problems of natural intentionality. For in the classical scientific image of the world as ruled by unbreakable natural laws, only a deity could possibly have created such "unnatural" beings as you and me.

References

Bruni, L. E. (2007). Cellular semiotics and signal transduction. In *Cellular Semiotics and Signal Transduction*, ed. Barbieri M. Dordrecht: Springer, 365–407.

Campbell, D. T. (1974). "Downward causation" in hierarchically organised biological systems. In *Downward Causation*, eds F. I. Ayala and T. Dobzhansky. Berkeley: University of California Press, 179–186.

Clayton, P., and Kauffman, S. (2006). On emergence, agency, and organization. *Biology and Philosophy*, 21: 501–521.

Deacon, T. (1997). *The Symbolic Species*. New York: Norton.

Deacon, T., and Sherman, J. (2008). The pattern which connects pleroma to creatura: The autocell bridge from physics to life. In *A Legacy for Living Systems: Gregory Bateson as Precursor to Biosemiotics*, ed. J. Hoffmeyer. Dordrecht: Springer, 59–76.

Deely, J. (2007). *Intentionality and Semiotics: A Story of Mutual Fecundation*. Scranton, PA: University of Scranton Press.

Depew, D. J., and Weber, B. H. (1995). *Darwinism Evolving: Systems Dynamics and the Genealogy of Natural Selection*. Cambridge, MA: Bradford/The MIT Press.

Gibbs, W. (2001). Cybernetic cells. *Scientific American*, August, 285: 52–57.

Gordon, D. M. (1995). The development of organization in an ant colony. *American Scientist*, 83: 50–57.

Gordon, D. (1999). *Ants at Work. How an Insect Society is Organized*. New York: The Free Press.

Gould, S. J. (2002). *The Structure of Evolutionary Theory*. Cambridge, MA: The Belknap Press of Harvard University Press.

Gray, R., Griffiths, P., and Oyama, S. (2001). *Cycles of Contingency. Developmental Systems and Evolution*. Cambridge, MA: MIT Press.

Griffiths, P. E., and Gray, R. D. (1994). Developmental systems and evolutionary explanation. *Journal of Philosophy*, 91 (6): 277–304.

Hoffmeyer, J. (1996). *Signs of Meaning in the Universe*. Bloomington, IN: Indiana University Press.

Hoffmeyer, J. (1998). Semiosis and biohistory: A reply. *Semiotica*, 120 (3/4): 455–482.

Hoffmeyer, J., and Kull, H. (2003). Baldwin and biosemiotics: What intelligence is for. In *Evolution and Learning*, eds B. Weber and D. J. Depew. Cambridge, MA: MIT Press, 253–272.

Hoffmeyer, J. (2008). *Biosemiotics. An Examination into the Signs of Life and the Life of Signs*. Scranton, PA: University of Scranton Press.

Kauffman, S. (1993). *Origins of Order: Self-Organization and Selection in Evolution*. New York: Oxford University Press.

Kauffman, S. A. (2000). *Investigations*. New York: Oxford University Press.

McFall-Ngai, M. J., and Ruby, E. G. (1998). Sepiolids and vibrios: When first they meet. *BioScience*, 48: 257–265.

Sarkar, S. (1996). Biological information: A skeptical look at some central dogmas of molecular biology. In *The Philosophy and History of Molecular Biology: New Perspectives*, ed. S. Sarkar. Dordrecht: Kluwer, 187–231.

Scriver, C. R., and Waters, P. J. (1999). Monogenic traits are not simple. *Trends in Genetics*, 15(7): 267–272.

Ulanowicz, R. E. (2009). *A Third Window: Natural Foundations for Life*. Philadelphia: Templeton Foundation Press.

Wilson, E. O. (1975). *Sociobiology. The New Synthesis*. London: Belknap Press.

11

Care on Earth: generating
informed concern

HOLMES ROLSTON, III

~

Evolutionary natural history has generated "caring" – by
elaborating, diversifying, conserving, and enriching such
capacities. A first response might be to take care about that
"caring"; the word is too anthropopathic. The framework
one expects in contemporary biology is rather termed the
evolution of "selfishness" (as if that word were not also
anthropopathic). Selfishness, however, is but one form of
caring; "caring" is the more inclusive term. Minimally,
biologists must concede that organisms survive and live
on, and that, over generations, they seek adapted fit. Or,
if "seek" is still too anthropopathic, they are selected
for their adapted fit. Maybe "select" is still too anthropopathic. Try computer language: the organic systems
are "calculating." Whatever the vocabulary, for all living
beings some things "make a difference"; they do not survive unless they attend to these things.

At least after sentience arises, neural organisms, human
or not, evidently "care." Animals hunt and howl, find
shelter, seek out their habitats and mates, feed their
young, flee from threats, grow hungry, thirsty, hot, tired,

Information and the Nature of Reality: From Physics to Metaphysics, eds. Paul Davies
and Niels Henrik Gregersen. Published by Cambridge University Press
© P. Davies and N. Gregersen 2010, 2014.

excited, sleepy. They suffer injury and lick their wounds. Sooner or later every biologist must concede that "care" is there. Call these "interests" or "preferences" or whatever; if "caring" is too loaded a term, then call these animal "concerns." Staying alive requires "self-defense." Living things have "needs." One of the hallmarks of life is that it can be "irritated." Organisms have to be "operational." Biology without "conservation" is death. Biology must be "pro-life." If you dislike the connotations of "caring," there are dozens of good biological terms that spiral around this term.

When humans arrive on the scene, "caring" is present by any conceivable standard. So once there was no caring; now "caring" is dominant on Earth. We need an account of its genesis. This will first be descriptive, but the description will demand prescription. Asking what humans and nonhumans *do* care about invites the question, for any who have choices, of what they *ought* to care about.

A consensus claim by those in complexity studies is that complex systems must be understood at multiple levels. Another is that simpler systems can generate more complex ones. Even so, neither complexity nor caring can be formalized into any comprehensive algorithmic routines. We will here rise through hierarchies and cross the thresholds requisite to the generation of caring. Complexity capable of caring, at least as we know it on Earth, is middle-range. Protons, electrons, and atoms do not care; nor do galaxies or stars. The human world, in which there is much caring, stands about midway between the infinitesimal and the immense on the natural scale. The greatest complexity we know is at our native range.

By some accounts, humans are dwarfed and shown to be trivial on the cosmic scale. By some accounts, humans are reduced and shown to be nothing but electronic molecules in motion on the atomic scale. But by equally impressive accounts, humans live at the center of complexity. In astronomical nature and micronature, at both ends of the spectrum of size, nature lacks the complexity that it demonstrates at the mesoscales, found in our native ranges on Earth. Perhaps we humans are cosmic dwarfs; perhaps we are molecular giants. But there is no denying our mid-scale complexity. We humans live neither at the range of the infinitely small, nor at that of the infinitely large, but we might well live at the range of the infinitely complex. We live at the range of the most caring; we ourselves might embody the most capacity for caring.

Initially, the evolution of caring on Earth requires the generation of complex chemistries, developing into enzymatic self-reproduction, developing into life with self-interest. But caring gets complicated, since selves are implicated. They eat each other, but equally they depend on each other (even on what they eat). They must reproduce themselves. Self-defense requires adapted fit; living things, and hence their cares, are webbed together in ecosystems. There will be "relations" – in today's fashionable term "networking."

Caring is self-contained only up to a point; after that it is caring "about" these relationships, the contacts and processes with which one is networked. It is caring about others: if only a predator caring to catch and eat prey; the prey caring to escape; both caring for their young. Caring will be matrixed and selective within such matrices. Networking will require distinctions, differential concerns.

In humans, there arise more inclusive forms of caring. Such wider vision requires even more complexity, a complex brain that can evaluate others not only in terms of helps and hurts, but also with concern for their health and integrity. This radically elaborates new levels of cultural information, and caring. Humans care about family, tribe, nation, careers, and ideational causes, such as biological science, French literature, or the Christian faith. Ethics shapes caring. In due course, humans alone on the planet can take a transcending overview of the whole – and care for life on Earth.

As good a description of Darwinian natural selection as any, and one that perhaps connects better than others with human culture, is that the story of life on Earth is of the generation and regeneration of caring. We here seek – we "care about" – a grander narrative of this informed concern. There is caring wherever there is "agency," wherever there is "motivation," where there is "locomotion," perhaps even where there are "motors." Science gives us a mechanistic universe, many say, claiming now to have shed anthropocentrism, forgetting that in its original etymology even a "machine" is "for" something. Axiologically, there is the generation of "norms." Cybernetically, there is the generation of "programs." Psychologically, there is the generation of "preferences." Ethically, there arise "duties," "responsibilities." All these concerns are our concern here. We need as full a story as we can get about caring on Earth. At length and in the end, we will seek the metaphysical and religious significance of this generation of caring.

11.1 Attraction: pacted matter

We have an umbrella word, "attraction," that is applied to everything from gravity to sexual appeal. Perhaps this makes the word too vague to be useful without further specification. But perhaps we can use it to launch our inquiry with the observation that right from the primordial beginnings, long before any processes we might term "caring" have evolved, even elemental matter is prone to pact itself, to gather itself into clumps. From one perspective, in the big bang, everything is flying apart in a universe continually expanding; but from another perspective, in the non-isomorphic universe that results, matter clumps into stars, into galaxies.

In some of these stars, all the heavier elements are forged: more pacting together of electrons, protons, neutrons. Four fundamental forces hold the world together: the strong nuclear force, the weak force, electromagnetism, and gravitation. These range over 40 orders of magnitude; some involve repulsion as well as attraction – the push as well as the pull is used to hold things together. The mix of forces is both remarkable and complex. Apparently, in this universe at least, these forces, and the particle masses and charges involved, have to be about what they are if matter is to become more complex, a prerequisite for anything still more complex developing.

Some of these stars explode again, but matter re-pacts itself, into planets. On (at least) one of these planets, complexity increases again by (so to speak) many more orders of magnitude. On Earth something we do want to call "caring" arises. This is no simple continuum but a complicated, diffracted, and punctuated story. We will be challenged to integrate it.

Beyond aggregation, matter is regularly spontaneously organizing, as when molecules and crystals form. In some situations, especially with a high flow of energy over matter, patterns may be produced at larger scales (Prigogine and Stengers, 1984). These patterns may further involve critical thresholds, often called self-organized criticality (Bak, 1996). Such processes are "automatic," sometimes called "self-organizing." Initially the "auto" should not be taken to posit a "self" but rather an innate principle of the spontaneous origination of order. Such features of matter are prerequisite to the later formation of proactive "selves."

11.2 Cybernetic nature: genes "for"

In wild nature, in view of entropy, only minimal levels of complexity can arise automatically. More complex biomolecules will not be reproduced often and reliably enough by spontaneous assembly. Advanced levels will require maintenance of the formed and functional structures and processes. Breakthroughs to new levels of complexity may be initially spontaneous, but their maintenance will require directed assembly and repair because otherwise the spontaneous breakdown rate will overwhelm the spontaneous construction rate. Spontaneous formation is a starter, but if you have to start from scratch each time, you do not get far. Natural history evolves first simply by "templating," but later on this becomes "instruction"; "forming" becomes "informing."

A major transition, launching endless possibilities in both complexity and in caring, happens with the origin of life, which, at least as we now inherit it, is always coupled with genetics. Genetic cybernetics is often said to be about "information," but it is not just information "about"; it is information "for." A genetic sequence has a potential for being an ancestor in an indefinitely long line of descendant genotype/phenotype reincarnations. The gene does not contain simply descriptive information "about" but prescriptive "for." The gene will be a gene "for" a trait because there has been natural selection "for" what it does contributing to adaptive fit. The preposition "for" saturates both natural selection and genetics. Traits get "selected for"; and the code "for" these gets simultaneously "selected for" in the genes and "mapped"; there is the genotype that records the know-how to make the functional structure and processes in the phenotype. In this sense, there is specified complexity.

Biologists may insist that in natural selection the mutation and shuffling process is blind, random. Some recent geneticists think it more probabilistic (Herring et al., 2006). In the results that this non-intentional process produces, however, genes do act directed toward a future, under construction. Unlike natural selection "for," wherever it shows up in genetics, there is a "telos" lurking in that "for." Biological functions are "teleonomic" (Mayr, 1988), contrasted with causation (including the "pacting" causation and automatic spontaneous organization) in physics and chemistry. Magmas crystallizing into rocks, and rivers flowing downhill have results, but no such "end."

Can genes care? Can genes generate care? Initially an answer is: genotypes cannot care, but some of the phenotypes they generate can. Genes cannot "intend" anything, any more than can the forces of natural selection operating on genes. Interestingly, however, some theoretical biologists and philosophers have begun using the term "intentional" as descriptive of biological information in genes. John Maynard Smith insists: "In biology, the use of informational terms implies intentionality" (Maynard Smith, 2000, p. 177). That word still has too much of a "deliberative" component for most users, but what is intended by "intentional" is this directed process, going back to the Latin *intendo*, with the sense of "stretch toward," or "aim at." Genes have both descriptive and prescriptive "aboutness"; they do stretch toward what they are about. Genetic information is "intentional" or "semantic" in this perspective, if it is for the purpose of ("about") producing a functional unit that does not yet exist. It is *teleosemantic*.

Where there is information being transmitted, there arises the possibility of mistakes, of error. The DNA, which is coded to make a certain amino acid sequence that will later fold into a protein segment, can be misread. If the reading frame gets shifted off the "correct" triplet sequence, then the "wrong" amino acids get specified, and the assemblage fails. There is "mismatch." Often there is machinery for "error-correction." This complexity must "discriminate." None of these ideas makes any sense in chemistry or physics, geology or meteorology. Atoms, crystals, rocks, and weather fronts do not "intend" anything and therefore cannot "err." A mere "cause" is pushy but not forward-facing. By contrast, a genetic code is a "code for" something, set for control of the upcoming molecules that it will participate in forming. There is proactive "intention" about the future.

Such "caring" requires considerable complexity in genetics. There is transcription, translation, signaling, messaging, copying, reading, coding, regulation, communication. Genes produce structural proteins, but many genes are regulatory. The whole idea of regulation is a precursor toward "caring," which makes no sense in the inanimate world. The interaction between structure and regulation has proved complex. Contemporary bioinformatics is limited by lack of computing power adequate for analysis of the complexity of the genetic sequences under study. The genetics must be complex because the proteins produced must be complex. One challenge is to produce widely variant proteins by ringing the changes on genetic sequences. Another is to produce highly stereospecific biomolecules that function to discriminate – "recognize" is the usual biologist's term – the needed required (or dangerous) resources: as when a hemoglobin, on account of its allosteric conformation, uses an iron molecule to bind oxygen and transport it from lung to cell.

11.3 Going concerns: organismic self-organization and self-defense

If you doubt that there is "caring" in the genotype, then move to the level of the phenotype. "Skin" is a sign of caring. Perhaps you wish to say with "bark" only that life has to be "protected." Biologists will take that term "protect" inside the skin to immune systems, and, for that matter, down to lipid bilayers. If you prefer scientifically more imposing words than "skin" or "bark", then use that of Humberto R. Maturana and Francisco J. Varela: "autopoiesis" (*autos*, "self," and *poiein*, "to produce") (Maturana and Varela, 1980). At this level in natural history, there has emerged a somatic self with know-how to protect it. If this isn't yet "caring," it is on the threshold of it; we are approaching another complex criticality.

Living things are self-maintaining systems. They grow; they are irritable in response to stimuli. They reproduce. They resist dying. They post a discrete (if also semipermeable) boundary between themselves and the rest of nature, and they assimilate environmental materials to their own needs. They gain and maintain internal order against the disordering tendencies of external nature. They keep recomposing themselves, whereas inanimate things run down, erode, and decompose. Life is a local countercurrent to entropy. Organisms pump out disorder. This self has to be maintained against entropy, or, if you like, against "aging." The constellation of these characteristics is nowhere found outside living things, although some of them can be mimicked or analogically extended to products designed by living systems. A crucial line is crossed when abiotic formations get transformed into loci of information. The *factors* come to include *actors* that exploit their environment.

Stuart Kauffman concludes a long study of the origins of order in evolutionary history:

Since Darwin, biologists have seen natural selection as virtually the sole source of that order. But Darwin could not have suspected the existence of self-organization, a recently discovered, innate property of some complex systems ... Selection has molded, but was not compelled to invent, the native coherence

of ontogeny, or biological development ... We may have begun
to understand evolution as the marriage of selection and self-
organization.   (Kauffman, 1991)

Evolution is a complex combinatorial optimization process.
                                    (Kauffman, 1993, p. 644)

In this generation of life, "pacting" reappears in biological
forms – first simply, with aggregating cells, that is, cells
sticking together, not unlike aggregating matter in physics
and chemistry; later there emerges integrated multicellular life. There is "organization," now concerted "self-
organization," the building of "organs." Although integrated and differentiated multicellularity (as contrasted
with microbial aggregation) did not arise for a long time
(over a billion years!), when it did arise, it arose more than
once (Bonner, 1998).

Complexity, as is often noticed, is modular. Within
organismic skin or bark, there are modules that need
boundaries, compartments, and walls, or at least boundary
zones. In living systems, a principal form of this modularity is cellular. There must be some means of preserving
modular identity. Distinction will require differentiation.
That makes things more complex and invites open-ended
differentiation, escalating the complexity. One measure of
complexity is the number of different cell types. Another
measure is the number of different kinds of interaction.
These modules in their nested sets must be reproduced
by directed reproduction. This level of information, as we
were saying, is proactive about assembly of these metabolic
modules and organs. Again, if this is not sufficient for caring, it is a necessary precursor.

There is not much point to modular complexity unless it enables more sophisticated information processing. An organism can survive better if it can exploit the beneficial regularities and deal with more of the contingencies in its environment than a competing organism. Organisms are "built to run, not fail." They are "over-engineered" with layers of control, selected for more complexity, more fail-safe protections, in the defense of their caring (Oliveri and Davidson, 2007). Perhaps this will be through tried and true stereotyped behaviors that win out statistically. But at least some organisms might better survive if there is flexibility; and this requires that an organism evaluate internal and external signals and respond by increasing or decreasing metabolisms and behaviors. The organism has to deal with nested sets of hierarchically organized proteins, lipids, enzymes, organelles, organs, an organismic system; a whole with its parts, surviving and reproducing in an ecosystem.

In modular systems, there will be systems and subsystems. Rarely will each module be equally and identically connected to each and all the other modules. Such hierarchical stratified systems must be described on multiple descriptive and interpretive levels. The construction of complexity begins with the simple and becomes more and more complex; the story is from the bottom up. But after certain thresholds are reached, parts can be differentiated. There is division of labor, a principal result of multicellularity. Often these relations are nonlinear; causes and effects are not proportionally related. There may be domains in which small changes in causal inputs make major changes in outputs, or vice versa: where huge changes in causal inputs make little change in outputs. If the modularity is evolving, the organic system will build by reiteration and modification of modules (mutations into something a little different), although there may also be novel sorts of modules, perhaps by co-option.

This requires more careful control of, and experimenting with, what is somatically inside the organism, but this is in response to external opportunity and threat. Modules interact not only with themselves, but also in and with an environment. Outside, the organism must also deal with thresholds and phase transitions, as when ice melts and there is water to drink. An epidemic microorganism might cause epidemic disease in a forest, and animals and birds must migrate to elsewhere on the landscape. Environments can be of multiple kinds, affecting the modular complexity. The environment of a complex biological organism is a mixedly ordered and chaotic mosaic with its variously predictable, probable, and indeterminate elements.

Like the ecologists who come later and study ecosystems to find that they resist law-like specification, organisms have evolved in environments that are both patterned and open. Much is local and site-specific. The courting display of the boreal owls differs from that of the ivory-billed woodpeckers, and each makes a certain sense in their niches in their environments. So? Rejoice in the particular and local forms of caring.

Such an Earth might first be thought to be a disvalue, but, on further investigation, if organisms operate in environments that are neither totally predictable on the one end of a spectrum, nor totally chaotic on the other, this generates a natural history that is much more "interesting" just because of its combining nomothetic and idiographic elements (to use the technical terms). If the world is perplexing, this stimulates more complex ways of dealing with it, and this perplexing–driving–complexing in turn demands more diverse caring. If some increment of capacity for innovative behavior appears, this potential will become more fully actualized in a challenging environment, with survival success. This will be repeated continually. Not only does living on such an Earth require organisms to generate more complexity in the parts, it also requires complex constructions to emerge that can oversee the behaviors of the parts; the top overtakes the bottom. So the ATP is being hydrolyzed to ADP, and the hemoglobin moves at an increased pace because the coyote is chasing the ground squirrel. And this is not just the nitrogen atoms in muscles and the iron atoms in hemoglobin, supporting the chase. The genes are being unzipped and read because the coyote needs more ADP. Genes are sometimes executive, as with the assembly of embryo. But when the organism has been constructed and is launched into ongoing metabolism, the phenotypic organism becomes equally executive. The organism uses its genes as a sort of Lego kit, in which it finds the assembler codes for the materials it comes to need.

Genetic behavior in and of itself is stereotyped, although these routines may be labile enough to respond to environmental stimuli. *Stentor roeseli*, a trumpet-shaped one-celled aquatic organism, has a mouth at the top and attaches itself by a foot to the substrate. If irritated, it contracts, or ducks, bending first this way and then that, or reverses the ciliary movement of its peristome to sweep water currents away. It may withdraw into a mucous tube about the base, to return after a few minutes, and, upon further irritation, repeat various avoidance reactions. But finally, with a jerk, it will break the attachment of its foot and swim away to attach itself elsewhere. If the irritation continues, it dislodges and tumbles away until by chance a non-noxious solution is found, whereupon it stops moving; the random locomotor variation ceases.

Stentor has no career memory; the particular *Stentor* cannot store previous solutions and invoke them at the next irritation. It cannot, for instance, invoke memory to "know" the next time to continue in the same direction as before as a direction of likely escape from irritation. There is no conditioned learning. Nevertheless, there is a sense in which this is intelligent and caring behavior, if also at a low level – from our human perspective. If irritated, *Stentor* generates and tests for more congenial places to live. There is probing and feedback. *Stentor* is doing this because it worked in previous *Stentors*, who survived, and this "intelligent" behavior got coded in the DNA. More complex organisms can climb a gradient or preferentially select food.

A still more sophisticated level of complexity, moving further toward caring, is reached with the capacity for learning, for acquired behavior. A coyote has a memory and conditioned learning; it can remember in which directions to run for cover. This requires developing neural or other capacities to operate in the subtleties of context, which in turn generates new levels of caring.

Such complexity involves emergence. The mutual interactions of the components and subsystems result in a capacity for behavior of the whole that transcends and is different from that in the parts and unknown in the previous levels of organization. Nursing evolves incrementally; and yet later on, there are mothers who nurse their young, while earlier there were none. They are genetically impelled to do so, but they also learn from watching other mothers nursing their young. They learn when and how to wean their young. Human mothers discipline and teach their young about what to eat and what not to eat. Now there is acquired learning.

Genes have to be understood from both a contemporary and a historical perspective. Why is the plant bending toward the sun? ("Orienting" itself, which we might think of as a precursor of caring.) Biologists need both proximate explanations and evolutionary explanations (Mayr, 1988, p. 28). Cells on the darker side of a stem elongate faster than cells on the brighter side because of an asymmetric distribution of auxin moving down from the shoot tip. But the explanation at a more comprehensive level is that, over evolutionary time, in the competition for sunlight, there were suitable mutations, and such phototropism increases photosynthesis.

So the explanation of this orienting stretches from the microbiology over a few hours to that of macroscale natural history over millennia. At more complex levels, there will be proximate explanations of how the mind shapes behavior, moving mammal mothers to nurse their children. At a more comprehensive explanatory level, this is because such mothers increased their number of offspring and were naturally selected. If you like, and to phrase it a little provocatively, there is natural selection for better caring. In cultures, this continues with cultural mores enabling one generation to produce and rear the next. The macrohistory drives the microhistory. The pro-life explanations can get prolific.

At this level, top-down and bottom-up issues arise. Where there is brain, this is as evidently top-down as the head is at the top of the body, directing its motions (as we later elaborate). These brained organisms do not behave the way they do simply on account of anything we know about electrons and protons, but on account of what we know about neural networks: indeed, on account of what we know about predators and prey, or, in human affairs, about trust funds and stock markets, or how to will an estate to the next generation.

Yes, when life appears, there is pro-life caring; but – hard-nosed biologists may now insist – the caring is entirely selfish. George C. Williams puts it bluntly: "Natural selection . . . can honestly be described as a process for maximizing short-sighted selfishness" (Williams, 1988, p. 385). Richard Dawkins has long insisted that every gene is a "selfish gene" (Dawkins, 1989). Here a principal worry, especially for any who have been apprehensive about my use of "caring," is that a moral word, "selfish," has been taken out of the human realm and applied to all living things, whether oysters or DNA molecules. At least, at the genetic level, biologists must concede that seldom, if ever, can single genes manage to be selfish, as genes must cooperate with each other in the whole organism, which survives or dies depending on this complex integration.

Since we are en route to an understanding of how "caring" has been generated, why not use a less pejorative word and say that each organism is in pursuit of – that is – values, its own *proper* life (from the Latin *proprius*: "one's own"), which is all that the (nonhuman) individual organism either can or ought to care to pursue? Bacteria, insects, crustaceans – including also the sentient creatures, the mice and chimpanzees – are projects of their own, each a life form to be defended for what it is intrinsically. An intrinsic value, from the perspective of biology, is found where there is a constructed, negentropic, cybernetic identity that is defended in such a somatic organismic self with an integrity of its own. Using its genes, the organism is acting "for its own sake," or, more philosophically put, "to protect its intrinsic value." These are "axiological genes."

Further, the increasing complexity is feeding the caring. Complexity, especially that which mixes order and openness, generates attention, preferential orientation. Such complexity produces perplexity. There arises both opportunity and threat. The organism can have fortunes and misfortunes. There can be cooperation and competition. The organism will have to generate and test better solutions to its problems. This escalates complexity, especially where there are competing cares that are matters of life and death. Biologists refer to this as coevolution, and notice that the dynamic ("the coevolutionary arms race," they may say) drives the generation of complexity. In this sense, a world with less chance could be a world with less caring.

11.4 Species concerns: (re)production of kindred others

Next, the horizon of caring enlarges, just because of the genetics. Even if biologists retain the "selfish" metaphor, the selfish caring gets "pacted" together, skin-out, as intensively as we saw it before skin-in. Not only must genes cooperate with each other in the whole organism, but the whole organism, the phenotype, gets placed in a species line. That first happens in reproduction: all species must "care" about reproduction – at least by Darwinian accounts that is their priority care – otherwise they become extinct.

So behaviors are selected that attend proportionately to the whole family: what biologists call "kin selection." One can still insist that the individual acts "selfishly" in his or her own interests, but "selfish" is now being stretched to cover benefits gained by "caring about" father, mother, niece, nephew, cousin, children, aunts, uncles, and so on, however far one chooses to look along the indefinitely extended lines of relationship, lines that fan out eventually to all conspecifics (half of which are also potential mates, which sometimes also need to be cared about). So it turns out that any such individual's "own proper life" is not exclusively individually owned, but is scattered about in the family, and that the individual competently defends its so-called "self" whenever possible and to the extent that this is manifested in the whole gene pool. This means that values about which the organism cares can be held intrinsically only as they are more inclusively distributed, and that places us in a position further to consider this process by which caring complexity is generated.

This proactive agency "cares" (still using that word provocatively) about the ongoing species line. Genera-tion requires regeneration. In reproduction, organisms reproduce themselves by passing a single set of minute coding sequences from one generation to the next, with the next generation self-organizing from this single trans-ferred information set. A single totipotent cell, using pro-vided maternal resources, transforms itself into a whole complex organism. What is conserved is not the matter, not the organism, not the somatic self, not even the genes, but a message that can only be conserved if – and only if – it is distributed, disseminated.

The passage of genes is the passage of species. The genes are the species writ small, the macroscopic species in microscopic code; the species is the genes writ large, the microscopic code in macroscopic species. The survival of the genes is the survival of the kind and vice versa, since genes code kind, and kind expresses genes. A genetic set codes the kind, representatively; and the organism, an expression of the kind, presents and represents the kind in the world.

Reproduction is typically assumed to be a need of individuals, but since any particular individual can flourish somatically without reproducing at all, indeed may be put through duress and risk or spend much energy reproducing, by another logic we can interpret reproduction as the species keeping up its own kind by reenacting itself again and again, individual after individual. It stays in place by its replacements. In this sense, a female grizzly bear does not bear cubs to be healthy herself, any more than a woman needs children to be healthy. Rather, her cubs are *Ursus arctos horribilis*, threatened by nonbeing, recreating itself by continuous performance. A species in reproduction defends its own kind. A female animal does not have mammary glands nor male animal testicles because the function of these is to preserve its own life; these organs are defending the line of life bigger than the somatic individual. The lineage in which an individual exists dynamically is something dynamically passing through it, as much as something it has. The locus of the intrinsic value – the value that is really defended over generations – seems as much in the form of life, the species, as in the individuals, since the individuals are genetically impelled to sacrifice themselves in the interests of reproducing their kind. What they "care about" is something dynamic to the specific form of life; they are selected to attend to the appropriate survival unit.

Selfish genes these reproducing individuals may have, but the genes "care more about" the species (so to speak) than the individual and its concerns. The solitary organism, living in the present, is born to lose; all that can be transmitted from past to future is its kind. Although selection operates on individuals, since it is always an individual that copes, selection is for the kind of coping that succeeds in copying, that is reproducing the kind, distributing the information coded in the gene more widely. Survival is through making others, who share the same valuable information. The organism contributes to the next generation all that it has to contribute, its own proper form of life, what it has achieved that is of value about how to live well its form of life. Survival is of the better sender of whatever is of genetic value in self into others. Survival of the fittest turns out to be survival of the senders. What genes are "for," we earlier said, is to be ancestors in an indefinitely long line of descendant genotype/phenotype reincarnations. Genes get "spread" around, or "distributed" by organisms who do not simply live for their "selves," but to spread what they know to other selves. Again, if doubters wonder whether this is yet to be called "caring," no doubt it is moving in that direction.

11.5 Natural selection: searching for adapted fit

On Jupiter and Mars, there is no natural selection. Nothing is competing, nothing is surviving, nothing has adapted fit. Nothing cares about anything. Even on Earth, climatological and geomorphological agitations continue in the Pleistocene period more or less like they did in the Precambrian. But the life story is different, because in biology, unlike physics, chemistry, geomorphology, or astronomy, something can be learned, first genetically and later neurally. The result, where once there were no species on Earth, there are today five to ten million. These species have been required to fit together in ecosystems, with adapted fit.

Complexity is modular, internally modular, as we earlier noticed from the skin in, but with adapted fit, complexity becomes also externally modular, from the skin out. Complexity is an organism-in-environment dialectic. Ecosystems have what the ecologists call niche space; rich ecosystems have myriads of such niches. A niche is both a place and a role in that place. Organisms make a living in a niche; they get webworked and fitted into trophic pyramids, into feedloops and feedback loops. They get figured into hypercycles, which are more complicated such loops. Upper levels depend on lower levels in these biotic communities, which means that the simpler forms of life, such as microbes and plants, do not vanish but remain essential in the pyramid of life.

This webworking remains essential, first, in the energy-capturing processes (animals would die without plants) but is also required in the material recycling processes (such as decay). Every living organism depends on value capture, often equally on feeding, whether grazing or predation, and on symbiosis. The biotic sector runs by need-driven individuals interacting with other such individuals and with the abiotic and exbiotic materials and forces (such as water and humus). Genotypes generate novel variations in phenotypes. The environment is not capricious, but neither is it regular enough to relax in. One always needs better detectors and strategies.

Dialectic with the loose environment (rich in opportunity, noisy, demanding in know-how) invites and requires creativity. The individual and the environment seem like *opposites*; they are really *apposites*; the individual is set opposed to its world but is also appropriate to it, in a niche, with adapted fit, faced with fortunes and misfortunes. The system has both order and openness, which invites more sophisticated caring. Situated environmental fitness often yields a complicated life together. Ecosystems are more or less stable, stable enough for natural selection to work dependably in them, and yet, equally often, the interactions are too messy to find the law-like regularities that scientists seek (Solé and Bascompte, 2006). Here ecologists may find such systems too complex to model effectively.

When such complexity builds up, life becomes something of an adventure. Living forms evolve in response to increasing niches in varied topographies, seen for instance in rapid and diverse insect speciation, resulting in highly specialized forms. Organisms must seek their niches. The mobile ones seek their habitats, seeking survival in these niches in which they are adapted. But again, this is not simply internal self-preservation, nor merely successful habitation of a niche. Selection places the individual on an ongoing species line, in a search for better adapted fit.

Natural selection, hard-nosed biologists may insist, does not "care." David Hume claimed that nature "has no more regard to good above ill than to heat above cold, or to drought above moisture, or to light above heavy" (Hume, 1972, p. 79). Or to life above nonlife, he would have added. That indifference can seem true from some perspectives, especially in the short term, although day-to-day nature is an impressive life support system. Sometimes it even seems true in the long term; every organism dies; species go extinct. But (as is always the case in complex systems) there is another perspective.

Nature on Earth has spun quite a story, making this planet with its landscapes, seascapes, and going from zero to five billion species in five billion years, evolving microbes into persons. Perhaps to say that nature "has regard" for life is the wrong way of phrasing it; we do not want to ascribe conscious caring to nature. Still, nature is a fountain of life. Nature is genesis. As a means to this genesis, natural selection demands fitness. Selection for adapted fit is a strange kind of indifference. Further, this fitness is not measured by an individual's own survival, long life, or welfare. Fitness is measured by what any individual can "contribute to" the next generation in its environment. Such fitness is not individualistic, not "selfish" at all; it is fitness in the flow of life, fitness to pass life on, to give something to others who come after.

Fitness is the ability to contribute more to the welfare of later-coming others of one's kind living in such niches, more relative to one's "competitors." That is an interesting way to think of natural selection: natural selection facilitates congruence between generations, selecting what genes, structures, and processes keep regenerating life in the midst of its perpetual perishing. In view of the larger religious horizons in which we are eventually interested, one could even employ a religious metaphor: fitness is "dying to self for newness of life" in a generation to come. Or, if you are not ready for that metaphor, still clinging to the metaphor of "indifference," then perhaps you have started to puzzle about when, during this long elaboration of life, "caring" ceases to be metaphorical and becomes literal.

11.6 Neural concerns: heading up felt experience

Your caring about caring is being done in the most complex object in the known universe: the human brain – and you just illustrated some of this complexity when you paused to puzzle over my preceding doubling-up of words: "caring about caring." Your primate brain, integrated with hands and legs, was generated as a survival tool in a "jungle" (or, more formally: a complex and chaotic world). Using instinct and conditioned behavior, lemurs "figure out" probabilities; there is that much order and contingency enough to churn the evolution of skills. This brain, in terms of our present interests, proved to be not only an impressive survival instrument; it did so by radically elaborating capacities for caring.

One of the most startling of such elaborations is the capacity for felt experience. The brain-object sponsors a subject with inwardness. Next after that, this subject/object brain elaborates its capacities for cognitive evaluation of the world that the embodied brain is moving through with such felt experiences. Now no one will deny that we have reached "caring" proper. At some critical level, complexity can become aware and, at a still higher threshold, self-aware. There emerges somatic self-awareness.

Animal brains are already impressive. In a cubic millimeter (about a pinhead) of mouse cortex there are 450 meters of dendrites and one to two kilometers of axons. The mouse brain is selectively organized; the mouse is interested in seed and ignores similar-looking pebbles. Interestingly, in terms of our interest here, although empathy is often thought to be unique to higher primates, possibly limited to humans alone, there is suggestive evidence even in mice that heightened sensitivity to pain occurs with the observation of familiar mice in pain. They seem to have brain enough for at least precursor capacities to sense what their fellow rodents are experiencing (Langford et al., 2006).

The human brain has a cortex 3000 times larger than that of the mouse. Our protein molecules are 97% identical to those in chimpanzees, differing by only 3%. But we have three times their cranial cortex, a 300% difference in the head. This cognitive development has come to a striking expression point in the hominid lines leading to *Homo sapiens*, going from about 300 to 1400 cubic centimeters of cranial capacity. The connecting fibers in a human brain, extended, would wrap around the Earth 40 times.

The human brain is of such complexity that descriptive numbers are astronomical and difficult to fathom. A typical estimate is 10^{12} neurons, each with several thousand synapses (possibly tens of thousands). Each neuron can "talk" to many others. The postsynaptic membrane contains over a thousand different proteins in the signal-receiving surface. "The most molecularly complex structure known [in the human body] is the postsynaptic side of the synapse," according to Seth Grant, a neuroscientist (quoted in Pennisi, 2006). Ever more intricate molecular interactions within synapses have made possible the circuitry that underlies our ability to think and to feel. These are "smart proteins." Over a hundred of these proteins were co-opted from previous, non-neural uses; but by far most of them evolved during brain evolution. "The postsynaptic complexes and the [signaling] systems have increased in complexity throughout evolution," says Berit Kerner, geneticist at the University of California, Los Angeles (quoted in Pennisi, 2006).

What is really exciting is that human intelligence is now "spirited," an ego with felt, self-reflective psychological inwardness. In the most organized structure in the universe, so far as is known, molecules, trillions of them, spin round in this astronomically complex webwork and generate the unified, centrally focused experience of mind. For this process neuroscience can as yet scarcely imagine a theory. A multiple net of billions of neurons objectively supports one unified mental subject, a singular center of concern and experience. Synapses, neurotransmitters, axon growth – all these can and must be viewed as objects from the "outside" when neuroscience studies them.

But what we also know, immediately, is that these events have "insides" to them: subjective experience. There emerges cognitive, existential, phenomenological self-awareness. There is "somebody there," already in the higher animals, but this becomes especially "spirited" in human persons (Russell et al., 1999). Very good evidence for this high-level spiritedness is found in how, when we persons make discoveries in neuroscience and genetics, this generates metaphysical and religious concerns. Our nearest relatives on Earth, such as the chimpanzees, do not come within a hundred orders of magnitude of such capacities, either for scientific cognition or for "spirited" (much less "spiritual") caring what to make of living on an Earth with such remarkable generative powers.

11.7 Minding concern: idea(l) commitments

All this "heads up" – with increased "top-down" causation – in self-reflective critical agents who have the capacity to build cumulative transmissible cultures. An information explosion gets pinpointed in humans. Humans alone have "a theory of mind"; they know that there are ideas in other minds, required for making these linguistic cultures possible. Such sophistication of language, grammar, and cognitive evaluation of felt and sought experiences brings increasingly complex caring. "Hundreds of millions of years of evolution have produced hundreds of thousands of species with brains, and tens of thousands with complex behavioral, perceptual, and learning abilities. Only one of these has ever wondered about its place in the world, because only one evolved the ability to do so" (Deacon, 1997, p. 21).

This cognitive network, formed and re-formed, makes possible virtually endless mental activity. The result of such combinatorial explosion is that the human brain is capable of forming more possible thoughts than there are atoms in the universe. We humans are the most sophisticated of known natural products. In our hundred and fifty pounds of protoplasm, in our three-pound brain is more operational organization than in the whole of the Andromeda galaxy. Some trans-genetic, nonlinear threshold seems to have been crossed. The geneticists reporting the sequencing of the human genome called this crossing a "massive singularity that by even the simplest of criteria made humans more complex" (Venter et al., 2001, p. 1347). All this activity is expressed in ever-elaborating forms of caring.

There is only one line that leads to persons, but in that line at least the steady growth of cranial capacity makes it difficult to think that intelligence is not being selected. "No organ in the history of life has grown faster" (Wilson, 1978, p. 87). One can first think that in humans enlarging brains are to be expected, since intelligence conveys obvious survival advantage. But then again, that is not so obvious, since all the other five million or so presently existing species survive well enough without advanced intelligence, as did all the other billions of species that have come and gone over the millennia. In only one of these myriads of species does a transmissible culture develop; and in this one it develops explosively, with radical innovations in cognition and caring that eventually have little to do with survival. Grigori Perelman sought and found a proof to the Poincaré conjecture in mathematics, transforming irregular spaces into uniform ones (Mackenzie, 2006). Edward O. Wilson cares for the conservation of his ants, where "splendor awaits in minute proportions" (1984, p. 139).

The power of ideas in human life is as baffling as ever. The nature and origins of language are proving to be, according to some experts in the field, "the hardest problem in science" (Christiansen and Kirby, 2003). Neuroscience went molecular (acetylcholine in synaptic junctions, voltage-gated potassium channels triggering synapsizing) to discover that what is really of interest is how these synaptic connections are configured by the information stored there, enabling function in the inhabited world. Our ideas and our practices configure and reconfigure our own sponsoring brain structures.

Thoughts in the conscious mind form and re-form, or, most accurately, in-form events in this brain space. We neuroimage blood brain flow to find that such thoughts can reshape the brains in which they arise. Our ideas and our practices configure and reconfigure our own sponsoring brain structures. In the vocabulary of neuroscience, we map brains to discover we have "mutable maps" (Merzenich, 2001, p. 418). For example, with the decision to play a violin well, and resolute practice, string musicians alter the structural configuration of their brains to facilitate fingering the strings with one arm and drawing the bow with the other (Elbert et al., 1995).

With the decision to become a taxi driver in London, and several years of experience driving about the city, drivers likewise alter their brain structures, devoting more space to navigation-related skills than have non-taxi drivers (Maguire et al., 2000). Similarly, researchers have found that "the structure of the human brain is altered by the experience of acquiring a second language" (Mechelli et al., 2004). Or by learning to juggle (Draganski et al., 2004). Or, we may as well suppose, by years of wondering how to solve the Poincaré conjecture, or how to classify and conserve the ants. The human brain is as open as it is wired up. No doubt our brains shape our minds, but also our minds shape our brains. The process is as top down as it is bottom up. Compare weather, a very complex system, with those resolute violin players remapping their brains.

Humans develop a discursive language in which words and texts become powerful symbols of the world, of the logic of that world, and of our place in the world. Humans have a double-level orienting system: one in the genes, shared with animals in considerable part; another in the mental world of ideas, as this flowers forth from mind, for which there is really no illuminating biological analog. We can now care about what is not at present seen (heard, tasted). There can be, so to speak, concern at a distance, caring not only interestedly but also disinterestedly about others. When knowledge becomes "ideational," these "ideas" make it possible to conceptualize and care about what is not present to felt experience. Chimps cannot care about the Ugandans in poverty, even if they encounter the poor at the edge of their forest, but Christians elsewhere in the world may, although they have never been to Uganda.

Conditioned learning among coyotes must take place in actual environmental encounter; but humans can imagine encounters, project them hypothetically, and learn from their imaginings. They have an idea-space, in their minds, which they can use as a trial-and-error simulator, and test in thought-experience behaviors that might gain what they care about. Such an idea-evaluator is faster and safer than trials conducted in the real world. The mental simulator can project the outcomes of such trials, and choose the best ones to test. Even the higher animals can do some of this, but human rationality enables humans to anticipate quite novel futures, to choose potential options, to plan for decades according to chosen simulations, or policies, and to rebuild their environments accordingly. The result is the capacity to care for idealized futures, and to work for such futures. Global capitalism is working now to make the rich richer and the poor poorer, but what if...?

11.8 Communitarian concern: tribalism and beyond

The result of these ideational powers – although persons continue to act in their generic self-interest – is to pull the focus of concern off self-center and bring into focus others in the community of persons. The rapid evolution of the human brain is driven not so much by the need for skills in tool-making or confronting the natural world as by the need to deal with social complexity, especially bonding cooperatively with others (Dunbar and Shultz, 2007). This drives the evolution of complex language, requiring complex and synaptically flexible brains. Bodily encounter presents us to each other; language with theory of mind "represents" (re-presents) each to the other. The single self must find a situated social fitness; a person ethically adapts to his or her neighbors. This produces community: initially tribalism. Tribalism – or, a little more inclusively, patriotism – is welcome, up to a point, but does caring evolve any further?

From the biological point of view, if natural selection operates on humans, it first seems plausible that those who care for self and family will out-reproduce those who care for any broader community. But individuals in their families are located within local communities, which, classically, have been tribes, bound together by their cultural mores. These mores can be beneficial when tribe encounters tribe. Group selection, long out of favor in biology, has been recently resurrected (Sober and Wilson, 1998). Tribes of "altruistic" cooperators will outreproduce tribes of selfish cooperators. Tribal ethics urges cooperating with kin and neighbors and defending the tribe against outsiders, often with the backing of tribal gods, or in sacred trust to ancestors. The benefits gained in out-group competition outweigh the costs of in-group cooperation. So we generate patriots in battle serving for God and country and the Rotarians building their public spirit.

But equally, Sober and Wilson insist, there is no "universal benevolence." "Group selection does provide a setting in which helping behavior directed at members of one's own group can evolve; however, it equally provides a context in which hurting individuals in other groups can be selectively advantageous. Group selection favors within-group niceness and between-group nastiness. Group selection does not abandon the idea of competition that forms the core of the theory of natural selection" (Sober and Wilson, 1998, p. 9).

Even from evolutionary theory, however, one can reach limited reciprocity with the competing out-groups. These others are not always enemies; they may be tribes with which we wish to trade. Or form alliances. A principal fact of modern life is increasingly wide, even global networks of reciprocity, evidenced in defense treaties and world trade, both cooperative and competitive. Here another threshold begins to be crossed, from tribalism to a more inclusive sense of community, and here Darwinian natural selection ceases to have sufficient explanatory power.

Beginning with a sense of one's own values to be defended, caring can sometimes become more "inclusive," recognizing that one's own self-values are widely paralleled, a kind of value that is distributed in myriads of other selves, in my tribe and in others. One comes to participate or share in this larger community of valued and valuing agents. The self-defense of value gets multiplied and divided by this interactive network of connections. The defense of one's own values gets mixed, willy-nilly, with the defense of the values of others.

Such acts can be understood in terms of conserving what the actors value, but the conservation of biologically based value underdetermines such events. The self is not simply biological and somatic, but cultural and ideological. What the self values can be sustained only if people act in concert. Cultural reproduction, conserving what one values in one's heritage, is as much required as is conserving one's genes. But much of one's cultural heritage is trans-tribal; one is drawn to the church catholic, to democracy, to a sense of fairness in international business, to conserving tropical forests.

All we had before was a concern for one's own advantage, but here "own advantage" has expanded over first to "shared advantage," "mutual advantage," not for all benefits but for many, which are shared first in the genetic kin and secondly in those who are axiologically kin, that is, who share one's culturally acquired values. These benefits often diffuse into those that cannot be differentially enjoyed – such as public safety or the right to vote, since what makes society safe and democratic for me (life pre-servers at the pool, free elections) confers these benefits at once on others. Trade works only if all keep promises, even across borders. With such benefits it is hard to be simply self-interested about them.

There will be trade-offs – my good against yours – and hence the sense of justice arises (each his or her due), or fairness (equitable outcomes for each), or of greatest good for greatest number. Such standards can appeal to every actor, in whatever culture (even though the detailed content will to some extent be culturally specific), because on the whole this is the best bargain that can be struck, mindful of the required reciprocation. Often it may be hard to reach more than a truce between parties pressing their self-interested cares, enlarged as these may be into kin, nations, corporations, and other reciprocating groups. In such disputes, issues of justice and fairness will arise.

A concern to behave fairly or justly is something more than a concern for self-interest, but at least those who press such self-interest publicly in national and international debates will have to do so in the name of fairness and justice. They will learn how to argue fairness and justice for their own sake, and perhaps will learn to feel the force of the unfairness and unjust allegations should these be used against them by others. If they cheat when they can get away with it, they themselves may realize that their conduct is unfair and unjust.

Further, there is considerable satisfaction both in being fairly and justly treated and in realizing that you keep your end of the bargain, even at some cost. What one ought to do, in any place, at any time, whoever one is, is what opti-mizes fairly shared values, and this is generically good, both for the self and the other, who are in parallel positions. One way of envisioning this is the so-called "original position," where one enters into contract, figuring out what is best for a person on average, oblivious to the specific circumstances of one's time and place, including one's genome and culture (Rawls, 1971).

Just this reflective element rationalizes (makes reasonable) and universalizes the recommended behavior. This is a sort of self-ignorant self-interest, where one is ignorant of all the particulars of one's self, and thereby must care about what would be generally in everybody's self-interest. The altruism is "indirect" in that there is no one-to-one benefactor-to-benefitted reciprocal exchange, but the altruism is quite "inclusive" just because of this indirectness. One expects to be helped out in a society of reciprocating helpers.

The problem with trying to cover this with a covertly selfish Darwinian explanation is that these indirect benefits are too pervasive. They loop back to the agent himself or herself, but they loop back to everybody else, nasty or nice, with about the same probability. They do not proportionately benefit the agent because they are benefiting both community and self. Natural selection cannot "see" the benefit to select any particular person's genome, because the agent, so far as that benefit is concerned, is not differentially benefited in producing more offspring.

This is where the sense of universality, or at least pan-culturalism, in morality has a plausible rational basis. Values must be recognized as widely dispersed, allocated, as having extensively proliferated beyond oneself; and now the protection of values has to be shared, some in self and some in others. Ethics develops into an effort to honor the intrinsic worth of persons, beginning with self and extending to others one encounters, and comes to require protecting them and what they value simultaneously with oneself. Toward their fellows, humans struggle with impressive, if also halting, success in an effort to evolve altruism in fit proportion to egoism.

11.9 Altruism: caring for others

Caring becomes more inclusive and more complex with the emergence of an inclusive altruism. The caring agent now has a world of concern, which has to be figured into self-concern and tribal concern. This takes increasing sophistication, partially because the determinants of concern that were previously so much in evidence, and which still continue, are now often transcended with an enlarging set of concerns. Nor does it seem that the scientific accounts, which had been comfortable enough with self-interests, are satisfactory in explaining this enlarging altruism.

Although one enters into the social contract in enlightened self-interest, morality can at times rise to still more enlightened consideration of the interests of others. After the agent has interiorized his or her bonding to others in society, he or she may come more and more to identify with those with whom values are shared. We have already seen how this can result in an equitable distribution of benefits and lies at the root of justice. But enlightened self-interest, supporting justice, is not the upper limit of moral development. Some persons, more than others, or all persons, some of the time more than other times, will move beyond such bargaining to envision a nobler humanity still to be gained in a more disinterested altruism that takes a deeper interest in others. We can consider and intend the interest of others, as part of our enlarged network of values in which the self is constituted.

This can motivate benevolence, beyond justice, where one acts to promote values respected in others, values both already there and facilitated by one's act of benevolence. One does not just fear loss from misbehaving others, but one is drawn to protect the benefits at stake in others by behaving morally toward them. Consider the Good Samaritan, with his expansive vision of who counts as a neighbor, a role model for millennia. Parallel models can be found in other traditions, as widespread variants of the Golden Rule illustrate. Concern for raising families, earlier enlarged into patriotism, now enlarges further into the concerned helping of non-genetically related, non-tribally related others. That there are, sometimes at least, Good Samaritans on Earth is as much a fact of the matter as is natural selection. These Samaritans, too, get regenerated generation after generation.

Biologists try to set such caring behavior in a Darwinian framework, but they find this challenging (Rolston, 2004). A tribe of such Samaritans would be likely to probably do well in competition with societies from which such behavior is absent. But this is not a tribal affair; this is cross-cultural concern. The determinant concern here is an "idea" (helping a neighbor) mixed with "felt experience" (sympathetic compassion) that jumps the genetics. Such ideational concern can be transmitted non-genetically, as has indeed happened in this case, since the story has been widely retold and praised as a model by persons in other cultures who are neither Jews nor Samaritans. The

Samaritan respects life not his own; he values life outside his own self-sector, outside his cultural sector. Neighbors are whomever one encounters that one is in a position to help.

By reductionist accounts within biology, these conscious altruistic motivations are superstructural, epiphenomenal. There are deep genetic determinants that generate the altruism, but these deep determinants must also simultaneously generate a superficial illusion of altruistic morality for the reciprocity to work sincerely enough. So the caring, although it seems to be present, is illusory. Cooperators frequently do well in society, and this works even better if the do-gooders (think they) are nobly motivated. The altruism is genuine at one level – the compassionate Samaritan believes he ought to help non-related others, and he is actively doing so – and illusory at another level. He is in fact behaving in such way as to increase the likelihood of his survival and of the survival of his children. He gains a good reputation; people think well of Good Samaritans and their children. What he really cares about is that people care about him. He is indeed helping the victim, but his larger concern is his own benefits.

Nor is this only disguised self-interest. A society of Samaritans sets up a climate of shared benefits. The Samaritan wishes to live in a world with many other Good Samaritans in it. I will help this victim; it is unlikely he will ever help me. But my helping sets up a caring climate, and some other Samaritan will help me when I fall victim. We all wish to live in such a world of "indirect reciprocity" (Alexander, 1987, p. 153; Nowak and Sigmund, 2005). With such benefits, however, it is hard to be self-interested about them in any individualistic or immediate sense. There is a feedback loop from single persons to society at large. The "unselfish" act of any particular individual benefits not only the person immediately assisted but, since it sets up a larger climate, benefits unspecified beneficiaries; and this common good promoted redounds to the benefit of the individual self.

That entwines the "self" with the community at large, and there is nothing problematic about finding that self-interest is sometimes interlocked with the common good. We might not want to call such concerns pure altruism, but it is certainly not pure selfishness. Why not say that in certain areas, like public safety, there are shared values? Notice also that an ethical dimension is beginning to emerge, for, although those entering into such a social contract stand to gain on average, they also acquire obligations to support this contract.

To insist on interpreting all such Samaritan caring, direct and indirect, in a covert Darwinian framework may be to miss a critical new turning point in the evolution of caring: the emergence of this "idea" become "ideal" – altruistic love. What seems quite evident is that now concerns are overleaping genes. Once concerns can pass from mind to mind, people do better with genes flexible enough to track the best ideas about caring, whether their blood kin launched these concerns or not. No one doubts that ideas jump genetic lines. One does not need Semitic genes to be a Christian, any more than Plato's genes to be a Platonist, or Einstein's genes to adopt the theory of relativity. The transmission process is neural, not genetic.

We rejoice in more widely shared concerns. Biologically speaking, the concern now is that the new adherents soon cease to have any genetic relationship to the proselytizers. The commitment that one has to make transcends one's genetics. Darwinians will claim, correctly, that all such sharing behavior creates a climate in which all those involved prosper owing to the reciprocity generated. But if in this Samaritan climate all persons benefit equally, then the differential survival benefit required at the core of natural selection has vanished. Darwinian accounts are no longer plausible. Those who catch on to the Samaritan sympathies prosper, whatever their genetics.

The benefits are impossible to keep local and in-group. This moral concern is analogous to learning to build fires. Fire-building does bring survival benefit to those who learn it for the first winter, but by the next winter the nearby tribes, watching through the bushes, have stolen the secret. Within a decade everybody knows how, and there is no longer any differential survival benefit. The Good Samaritan ideal, like fire-building, has circumnavigated the globe, except that it is no stolen secret: it is spread by missionaries. This appearance of universalist creeds with their capacities to generate this more inclusive caring – even if more ideal than real – is one of the most remarkable forms of caring on Earth. "Do to others as you would have them do to you" helps us to cope because here is insight not just for the tribe, but for the world; indeed, if there are moral agents with values at stake in other worlds, this could be universal truth.

11.10 Altruism: caring for Earth

Once there was no caring on Earth; today caring has gone global – at least in ideal; sometimes, if but partly, in real. Ethics has been around for millennia; the Golden Rule is perennial. But we have only recently become aware of evolutionary natural history and threats to the biodiversity it has generated. Concern on Earth comes to include concern about Earth. This starts with human concerns for a quality environment, and some think this shapes all our concerns about nature from start to finish. Humans are the only self-reflective, deliberative moral agents. Ethics is for people. But humans co-inhabit Earth with five to ten million species. If the values that nature has achieved over evolutionary time are at stake, then ought not humans to find nature sometimes morally considerable in itself?

Nature has equipped *Homo sapiens*, the wise species, with a conscience. Perhaps conscience is less wisely used than it ought to be when, as in classical Enlightenment ethics, it excludes the global community of life from consideration, with the resulting paradox that the self-consciously moral species acts only in its collective self-interest toward all the rest. Perhaps we humans are not so "enlightened" as once supposed – not until we reach a more considerate, environmental, Earth ethic.

Several billion years' worth of creative toil, several million species of teeming life, have been handed over to the care of this late-coming species in which mind has flowered and morals have emerged. Ought not those of this sole moral species to do something less self-interested than count all the produce of an evolutionary ecosystem resource to be valued only for the benefits they bring? Such an attitude hardly seems biologically informed, much less ethically adequate. Its logic is too provincial for moral humanity. Such anthropocentrism is insufficiently caring.

Contemporary ethics has been concerned to be inclusive. Environmental ethics is even more inclusive. It is not simply what a society does to its slaves, women, black people, minorities, disabled, children, or future generations, but what it does to its fauna, flora, species, ecosystems, and landscapes that reveals the character of that society. Whales slaughtered, wolves extirpated, whooping cranes and their habitats disrupted, ancient forests cut, Earth threatened by global warming – these are ethical questions intrinsically, owing to values destroyed in nature, as well as also instrumentally, owing to human resources jeopardized. Humans need to include nature in their ethics; humans need to include themselves in nature.

Here is another critical threshold, recognizing a difference crucial for understanding the human possibilities in the world. Humans can not only be altruists one to another; they can be still more inclusive altruists when they recognize the claims of nonhuman others: animals, plants, species, ecosystems, a global biotic community. This most altruistic form of ethics embodies the most comprehensive caring. It really loves others. This ultimate altruism is, or ought to be, the human genius. In this sense the last becomes the first; this late-coming species with a latterday ethics is the first to see and care about the story that is taking place. This late species, just because it has the most capacity for caring, must take a leading role.

11.11 Logos and love

Einstein was impressed with the logic in the world and the human capacity to track this logic. "The eternal mystery of the world is its comprehensibility... The fact that it is comprehensible is a miracle" (Einstein, 1970, p. 61).

We can complement Einstein: The eternal mystery is that this world, generating rationality sufficient for worldviewing, has also generated capacities for caring necessary for appropriate respect in and of this world. Humans are "the rational animal"; that has been the classical philosophical claim. But much recent psychology and cognitive science have insisted also that humans, with their minds, are embodied, maintaining life, ever vigilant toward helps and hurts, both skin-in and skin-out. The mind is an instrument of both reasoned and unreasoned caring.

On another front, however, we might correct Einstein. "Our experience...justifies us in believing that nature is the realisation of the simplest conceivable mathematical ideas" (Einstein, 1934, p. 36). On the contrary, nature seems bent on becoming more complex. If this is not universally true, it is at least true of Earth's natural history. The world has never yet proved as simple as we thought. In the real worlds of both nature and culture, it seems unlikely that there is any calculus of discovery, any set of formal operations that is universalizable, no general problem solver (GPS), no general or synthetic systems theory. Perhaps it is remarkable that nature has generated a human mind complex enough to discover the mathematical simplicity in theoretical physics. But the generating of such a mind has required a complex and diverse environment.

Such intelligence evolves to operate in a mixedly patterned and open world. Formal operations are only a part of such intelligence. Much complexity research has been based on computer models, with a certain paradox: that the computer models, although they may be supposed to "model" the real world, of necessity abstract from a more complex real world. They "simulate"; they are simulacra. A central puzzle here is that, although computers can "simulate" care (be programmed to maximize this or that value), they cannot in fact "care." Organisms live and die; computers do not. Organisms reproduce their species lines; computers are artifacts, not a natural kind. Organisms are naturally selected, contemporary organisms the outcome of three-and-a-half billion years of evolutionary natural history. Computers are not yet a century old; nor do they reproduce themselves.

Brains rapidly evolved, as we saw earlier. Such evolved intelligence, in contrast to mechanized intelligence, can and must come to care about ongoing history, and its role in it. We may desire, for instance, to preserve and enlarge family and tribe. We may come to care that democracy survives in the world, or that the wisdom of Shakespeare not be lost in the next generation, and to work to fulfill such ideals. Perhaps computers will be built that can think (play chess). But if they cannot also emote, desire, weep, love, make decisions involving free will, if they have no "affect," then they cannot do the sort of caring-thinking related to these psychological states. Intelligence includes, for example, the ability to discern analogies and parallels between outwardly dissimilar phenomena, to disambiguate equivocations. It includes the capacity to detect gestalts, to follow developing story lines. We might construct a computer with the capacity to search out mathematical simplicity, but what would it be like to construct a computer with a good sense of plot, caring about how the story ends? The embodied mind is not hardware, not software; it is (so to speak) wetware that must be kept wet, sometimes with tears.

We must attach logic to loving, if we are to understand what natural history has done. Reason is yoked with emotion, cognition with caring. So far from resisting this, let us welcome it. But as we welcome this connection, we have also to recognize that science, whatever else it discovers for us about the evolution of caring, is incompetent henceforth to direct it. When we yoke logic and love, we reach yet another critical threshold: the gap between the *is* and the *ought*. For that one needs ethics. Ethics is the choice of the right, in the face of temptation to do otherwise. "Temptation" is poorly embodied in computers, almost as poorly as right and wrong is addressed in science. The natural forces, thrusting up the myriad species, produced one that, so to speak, reached escape velocity, transcending the merely natural with cares super to anything previously natural. A complement of this eternal mystery is the possibility for better and worse caring, for noble and for misplaced caring, for good and evil. Humans are capable of pride, avarice, flattery, adulation, courage, charity, forgiveness, prayer.

R. L. Stevenson pondered the "incredible properties" of dust stirring to give rise to this creature struggling for responsible caring:

What a monstrous spectre is this man, the disease of the agglutinated dust, lifting alternate feet or lying drugged with slumber; killing, feeding, growing, bringing forth small copies of himself; grown upon with hair like grass, fitted with eyes that move and glitter in his face; a thing to set children screaming; – and yet looked at nearlier, known as his fellows know him, how surprising are his attributes! Poor soul, here for so little, cast among so many hardships, filled with desires so incommensurate and so inconsistent, savagely surrounded, savagely descended, irremediably condemned to prey upon his fellow lives: who should have blamed him had he been of a piece with his destiny and a being merely barbarous? And we look and behold him instead filled with imperfect virtues: infinitely childish, often admirably valiant, often touchingly kind; sitting down, amidst his momentary life, to debate of right and wrong and the attributes of the deity; rising up to do battle for an egg or die for an idea; singling out his friends and his mate with cordial affection; bringing forth in pain, rearing with long-suffering solicitude, his young. To touch the heart of his mystery, we find in him one thought, strange to the point of lunacy: the thought of duty; the thought of something owing to himself, to his neighbour, to his God: an ideal of decency, to which he would rise if it were possible; a limit of shame, below which, if it be possible, he will not stoop.   (Stevenson, 1903, pp. 291–295)

The embodied story is the human legacy of waking up to good and evil (as in Genesis 1–2), or the dreams of hope for the future (as with visions of the kingdom of God). This, as much as logic and love, may be the *differentia* of the human genius. The generation of such caring is as revealing as anything else we know about natural history. The fact of the matter is that evolution has generated ideals in caring.

Nor should we be surprised that this generating has been a long struggle. The evolutionary picture is of nature laboring in travail. The root idea in the English word "nature," going back to Latin and Greek origins, is that of "giving birth." Birthing is creative genesis, which certainly characterizes evolutionary nature. Birthing (as every mother knows) involves struggle. Earth slays her children, a seeming evil, but bears an annual crop in their stead. The "birthing" is nature's orderly self-assembling of new creatures amidst this perpetual perishing. Life is ever "conserved," as biologists might say; life is perpetually "redeemed," as theologians might say. From our perspective, let us call it the "generation and regeneration of caring."

Perhaps the planetary set-up is an accident, but the ongoing processes after the set-up seem to be loaded with fertility. Life depends on a statistical stability blended with open contingency; in the short term all lose, death is inevitable; but then again, in the long term life persists, phoenix-like, in the midst of its destruction. There is a kind of "promise" in nature, not only in the sense of potential that is promising, but also in the sense of reliability in the Earthen set-up that is right for life. Perhaps nature does not care, but nature is a care-generator, since nature does evolve, over the millennia, billions of species in almost every nook and cranny of Earth in which caring for life does take place.

Complex systems, we have been saying, have to be understood at multiple levels. So move to a different perspective on this Earthen story. This churn of materials, perpetually agitated and irradiated with energy, is not only to be seen as indifferent resource but as prolific source. The negentropy is as objectively there as the entropy, the achievements as real as the drifting cycles and random walks. Against the indifference, we now must counter that the systemic results have been prolific, today five million species flourishing in myriads of diverse ecosystems. One species is challenged to care for the whole community of life, challenged today more than ever before in human history. To say that there is nothing but systemic indifference seems to ignore these principal results of natural history, including those embodied in us. Even those who retain doubts about natural systems cannot doubt that in human systems caring is omnipresent. Nor that better caring is urgent and among our most challenging tasks.

Dealing with causes, we interpret the results in terms of the precedents (A causes B). Dealing with stories, and histories, however, we may need to interpret the beginnings by thinking back from the endings (Y has been unfolding toward Z). Complexity is often to be understood not just bottom up, but top down. To that we add, in closing, that the caring-complexity in which we find ourselves must be understood comprehensively – in terms of conclusions, not just origins. That ending lies ahead, but en route, we humans are at the forefront of the story. Increased caring, like the increased complexity that supports it, is an ever open niche. That invites us to see such a world, and our task in it, as sacred, even divine.

References

Alexander, R. D. (1987). *The Biology of Moral Systems*. New York: Aldine de Gruyter.

Bak, P. (1996). *How Nature Works: The Science of Self-Organized Criticality*. New York: Springer-Verlag.

Bonner, J. T. (1998). The origins of multicellularity. *Integrative Biology*, 1: 27–36.

Christiansen, M. H., and Kirby, S. (2003). Language evolution: The hardest problem in science? In *Language Evolution*, eds M. H. Christiansen and S. Kirby. New York: Oxford University Press, 1–15.

Dawkins, R. (1989). *The Selfish Gene*, new ed. New York: Oxford University Press.

Deacon, T. W. (1997). *The Symbolic Species: The Co-Evolution of Language and the Brain*. New York: Norton.

Draganski, B., et al. (2004). Neuroplasticity: changes in grey matter induced by training. *Nature*, 427: 311–312.

Dunbar, R. I. M., and Shultz, S. (2007). Evolution in the social brain. *Science*, 317: 1344–1347.

Einstein, A. (1934). On the method of theoretical physics. In *The World as I See It*. New York: Covici–Friede Publishers, 30–40.

Einstein, A. (1970). *Out of My Later Years*. Westport, CT: Greenwood Press.

Elbert, T., et al. (1995). Increased cortical representation of the fingers of the left hand in string players. *Science*, 270: 305–307.

Herring, C. D., et al. (2006). Comparative genome sequencing of *Escherichia coli* allows observation of bacterial evolution on a laboratory timescale. *Nature Genetics*, 38: 1406–1412.

Hume, D. (1972). *Dialogues Concerning Natural Religion*, ed. H. D. Aiken. New York: Hafner Publishing.

Kauffman, S. A. (1991). Antichaos and adaptation. *Scientific American*, 265 (no. 2): 78–84.

Kauffman, S. A. (1993). *The Origins of Order: Self-Organization and Selection in Evolution*. New York: Oxford University Press.

Langford, D. J., et al. (2006). Social modulation of pain as evidence for empathy in mice. *Science*, 312: 1967–1970.

Mackenzie, D. (2006). The Poincaré conjecture proved. *Science*, 314: 1848–1849.

Maguire, E. A., et al. (2000). Navigation-related structural change in the hippocampi of taxi drivers. *Proceedings of the National Academy of Sciences of the United States of America*, 97(8): 4398–4403.

Maturana, H. R., and Varela, F. J. (1980). *Autopoiesis and Cognition: The Realization of the Living*. Dordrecht, Boston: D. Reidel Publishing.

Maynard Smith, J. (2000). The concept of information in biology. *Philosophy of Science*, 67: 177–194.

Mayr, E. (1988). *Toward a New Philosophy of Biology*. Cambridge, MA: Harvard University Press.

Mechelli, A., et al. (2004). Neurolinguistics: structural plasticity in the bilingual brain. *Nature*, 431: 757.

Merzenich, M. A. (2001). The power of mutable maps. In *Neuroscience: Exploring the Brain*, 2nd ed., eds M. F. Bear, B. W. Connors, and M. A. Paradiso. Baltimore: Lippincott Williams and Wilkins, 418–452.

Nowak, M. A., and Sigmund, K. (2005). Evolution of indirect reciprocity. *Nature*, 437: 1291–1298.

Oliveri, P., and Davidson, E. H. (2007). Built to run, not fail. *Science*, 315: 1510–1511.

Pennisi, E. (2006). Brain evolution on the far side. *Science*, 314: 244–245.

Prigogine, I., and Stengers, I. (1984). *Order out of Chaos: Man's New Dialogue with Nature*. New York: Bantam Books.

Rawls, J. (1971). *A Theory of Justice*. Cambridge, MA: Harvard University Press.

Rolston, H. (2004). The Good Samaritan and his genes. In *Evolution and Ethics: Human Morality in Biological and Religious Perspective*, eds P. Clayton and J. Schloss. Grand Rapids, MI: William B. Eerdmans Publishing, 238–252.

Russell, R. J., et al., eds. (1999). *The Neurosciences and the Person: Scientific Perspectives on Divine Action*. Berkeley, CA: Center for Theology and the Natural Sciences.

Sober, E., and Wilson, D. S. (1998). *Unto Others: The Evolution and Psychology of Unselfish Behavior*. Cambridge, MA: Harvard University Press.

Solé, R. V., and Bascompte, J. (2006). *Self-Organization in Complex Ecosystems*. Princeton: Princeton University Press.

Stevenson, R. L. (1903). Pulvis et umbra. In *Across the Plains*. New York: Charles Scribner's Sons, 289–301.

Venter, J. C., et al. (2001). The sequence of the human genome. *Science*, 291: 1304–1351.

Williams, G. C. (1988). Huxley's evolution and ethics in socio-biological perspective. *Zygon*, 23: 383–407.

Wilson, E. O. (1978). *On Human Nature*. Cambridge, MA: Harvard University Press.

Wilson, E. O. (1984). *Biophilia*. Cambridge, MA: Harvard University Press.

PART IV
PHILOSOPHY AND THEOLOGY

~

12

The sciences of complexity: a new theological resource?

ARTHUR PEACOCKE

∼

A host of surveys indicate that what Christians, and indeed other religious believers, today affirm as 'real' fails to generate any conviction among many of those who seek spiritual insight and who continue regretfully as wistful agnostics in relation to the formulations of traditional religions – notably Christianity in Europe, and in intellectual circles in the USA. Many factors contribute to this state of affairs, but one of these, I would suggest, is that the traditional language in which much Christian theology, certainly in its Western form, has been and is cast is so saturated with terms that have a supernatural reference and colour that a culture accustomed to think in naturalistic terms, conditioned by the power and prestige of the natural sciences, finds it increasingly difficult to attribute any plausibility to it. Be that as it may, there is clearly a pressing need to describe the realities that Christian belief wishes to articulate in terms that can make sense to that culture without reducing its content to insignificance.

Correspondingly, there is also a perennial pressure, even among those not given to any form of traditional

Information and the Nature of Reality: From Physics to Metaphysics, eds. Paul Davies and Niels Henrik Gregersen. Published by Cambridge University Press © P. Davies and N. Gregersen 2010, 2014.

religiosity, to integrate the understandings of the natural world afforded by the sciences with very real, 'spiritual' experiences, which include interactions with other people and awareness of the transcendent.

Both of these pressures in contemporary life accentuate the need to find ways of integrating 'talk about God' – that is, theology – with the world view engendered and warranted by the natural sciences. The 'god of the gaps' and the whole notion of divine 'revelation' are inevitably expelled from such a world view; however, it will be argued that this is not equivalent to the expulsion of the creator God of the monotheistic religions.

Meanwhile, the scientific world view has itself not been static and has had to come to terms more and more with, inter alia, the ability of natural complexes to manifest new properties not associated with the components. Such complexes display, for example, self-organizing properties, and awareness of this and of the general role of the transfer of information in complex systems has generated a metaphysics to take account of this feature of the natural world: namely, 'emergentist monism' (see section 12.1, below). Furthermore, there has been an impetus under pressure from the scientific world view to revise ideas concerning God's relation to the world, described so differently from the past, in terms both of a 'theistic naturalism' (see section 12.2) and 'panentheism' (see section 12.3). The aim of this chapter is to integrate these developments and to show that this process allows one to formulate the role of theological language in such a way as to escape the impasse described above and thereby to justify its claim to refer realistically to the relationships of nature, persons, and God, which are the concern of religious experience and are the focus of intellectual reflection on it: namely, theology. The following reproduces here, for convenience, some previous expositions of mine on these three pertinent themes.[1]

12.1 Emergentist monism

More and more, the natural and human sciences give us a picture of the world as consisting of a complex hierarchy (or hierarchies) – a series of levels of organization and matter in which each successive member of the series is a whole constituted of parts preceding it in the series.[2] The wholes are organized systems of parts that are dynamically and spatially interrelated – a feature sometimes called a 'mereological' relation. Furthermore, all properties also result, directly in isolation or indirectly in larger patterns, from the properties of microphysical entities. This feature of the world is now widely recognized to be of significance in relating our knowledge of its various levels of complexity – that is, the sciences that correspond to these levels.[3] It also corresponds not only to the world

---

[1] The account of these developments (in sections 12.1, 12.2, 12.3) given here follows closely the expositions in my article, 'Emergent realities with causal efficacy' (Peacocke, 2007) (for section 12.1); and those in my *Paths from Science towards God* (Peacocke, 2001) (for sections 12.2 and 12.3).

[2] Conventionally said to run from the 'lower', less complex, to the 'higher', more complex systems, from parts to wholes, so that these wholes themselves constitute parts of more complex entities – rather like a series of Russian dolls. In the complex systems I have in mind here, the parts retain their identity and properties as isolated individual entities (Peacocke, 1994).

[3] See, for example, Peacocke (1993), especially Figure 1 (p. 195), based on a scheme of W. Bechtel and A. Abrahamson (1991, Figure 8.1).

in its present condition, but also to the way complex systems have evolved in time out of earlier simpler ones.

I shall presume at least this with the 'physicalists': all concrete particulars in the world (including human beings), with all of their properties, are constituted of fundamental physical entities of matter/energy manifest in many layers of complexity – a 'layered' physicalism. This is indeed a *monistic* view (a constitutively ontologically reductionist one) that everything can be broken down into whatever physicists deem to constitute matter/energy. No extra *entities or forces*, other than the basic four forces of physics, are to be deemed to be inserted at higher levels of complexity in order to account for their properties. However, what is significant about natural processes and about the relation of complex systems to their constituents now is that the concepts needed to describe and understand – as indeed also the methods needed to investigate each level in the hierarchy of complexity – are specific to and distinctive of those levels. It is very often the case (but not always) that the properties, concepts, and explanations used to describe the higher-level wholes are not logically reducible to those used to describe their constituent parts, themselves often also constituted of yet smaller entities. This is an epistemological assertion of a nonreductionist kind.

When the epistemological nonreducibility of properties, concepts, and explanations applicable to higher levels of complexity is well established, their employment in scientific discourse can often, *but not in all cases*, lead to a putative and then to an increasingly confident attribution of reality to that to which the higher-level terms refer. 'Reality' is not confined to the physicochemical alone. One

318

must accept a certain 'robustness'[4] of the entities postulated, or rather discovered, at different levels and resist any attempts to regard them as less real in comparison with some favoured lower level of 'reality'. Each level has to be regarded as a cut through the totality of reality, if you like, in the sense that we have to take account of its mode of operation at that level. New and distinctive kinds of realities at the higher levels of complexity may properly be said to have *emerged*. This can occur with respect either to moving, synchronically, up the ladder of complexity or, diachronically, through cosmic and biological evolutionary history.

Much of the discussion of reductionism has concentrated upon the relation between already established theories pertinent to different levels. This way of examining the question of reductionism is less appropriate when the context is that of the biological and social sciences, for which knowledge hardly ever resides in theories with distinctive 'laws'. In these sciences, what is sought is more usually a *model* of a complex system that explicates how its components interact to produce the properties and behaviour of the whole system – organelle, cell, multicellular organism, ecosystem, etc. These models are not presented as sentences involving terms that might be translated into lower-level terms for reduction to be successful but, rather, as visual systems, structures, or maps, representing multiple interactions and connecting pathways of causality and determinative influences

---

4 W. C. Wimsatt has elaborated this criterion of 'robustness' for such attributions of reality to emergent properties at the higher levels (Wimsatt, 1981).

between entities and processes. When the systems are not simply aggregates of similar units, then it can turn out that the behaviour of the system is due principally, sometimes entirely, to the distinctive way its parts are put together – which is what models attempt to make clear. This incorporation into a system constrains the behaviour of the parts and can lead to behaviour of the systems as a whole that is often unexpected and unpredicted (Richardson, 1992). As W. Bechtel and R. C. Richardson have expressed it: 'They are *emergent* in that we did not anticipate the properties exhibited by the whole system given what we knew of the parts' (Bechtel and Richardson, 1992, pp. 266–267). They illustrate this from a historical examination of the controversies over yeast fermentation of glucose and oxidative phosphorylation. What is crucial here is, not so much the unpredictability, but the inadequacy of explanation if only the parts are focused upon, rather than the whole system. 'With emergent phenomena, it is the interactive organization, rather than the component behaviour, that is the critical explanatory feature' (ibid., p. 285).

There are, therefore, good grounds for utilizing the concept of 'emergence' in our interpretation of naturally occurring, hierarchical, complex systems constituted of parts that themselves are, at the lowest level, made up of the basic units of the physical world. I shall denote[5] this position as that of *emergentist monism*.

---

[5] As does Philip Clayton, the author of Chapter 3 in this volume. Note that the term 'monism' is emphatically *not* intended (as is apparent from the nonreductive approach adopted here) in the sense in which it is taken to mean that physics will eventually explain everything (which is what 'physicalism' is usually taken to mean).

If we do make such an ontological commitment about the reality of the 'emergent' whole of a given total system, the question then arises of how one is to explicate the relation between the state of the whole and the behaviour of parts of that system at the micro level. It transpires that extending and enriching the notion of causality now becomes necessary because of new insights into the way complex systems, in general, and biological ones, in particular, behave.

A more substantial ground for attributing reality to higher-level properties and the organized entities associated with them is the possession of any distinctive causal (I would say, rather, 'determinative') efficacy of the complex wholes that has the effect of making the separated, constituent parts behave in ways they would not do if they were not part of that particular complex system (that is, in the absence of the interactions that constitute that system). For *to be real is to have causal power*.[6] New causal powers and properties can then properly be said to have *emerged* when this is so.

Subtler understanding of how higher levels influence the lower levels allows application in this context of the notion of a determining ('causal') relation from whole to part (of system to constituent) – never ignoring, of course, the 'bottom-up' effects of parts on the wholes, which depend on their properties for the parts being what they are, albeit now in the new, holistic, complex, interacting configurations of that whole. A number of related concepts have in recent years been developed to describe these

---

[6] A dictum attributed to S. Alexander by J. Kim (Kim, 1992, pp. 134–135; 1993, p. 204).

relations in both synchronic and diachronic systems – that is, both those in some kind of steady state with stable characteristic emergent features of the whole and those that display an emergence of new features in the course of time.

In particular, the term 'downward causation' or 'top-down causation' was employed by Donald Campbell (1974) to denote the way in which the network of an organism's relationships to its environment and its behaviour patterns together determine over the course of time the actual DNA sequences at the molecular level present in an evolved organism – even though, from the 'bottom-up' viewpoint of that organism once in existence, a molecular biologist would tend to describe its form and behaviour as a consequence of those same DNA sequences. Other systems could be cited,[7] such as the Bénard and certain autocatalytic reaction systems (for example, the famous Zhabotinsky reaction and glycolysis in yeast extracts), that display spontaneously, often after a time interval from the point when first mixed, rhythmic temporal and spatial patterns, the forms of which can even depend on the size of the containing vessel. Harold Morowitz (2002) has indeed identified some 28 emergent levels in the natural world. Many examples are now known also of dissipative systems which, because they are open, a long way from equilibrium, and nonlinear in certain essential relationships between fluxes and forces, can display large-scale patterns in spite of random motions of the units – 'order out of chaos', as Prigogine and Stengers (1984) dubbed it.

In these examples, the ordinary physicochemical account of the interactions at the micro level of description

---

[7] For a survey with references, see Peacocke (1983/1989).

simply cannot account for these phenomena. It is clear that what the parts (molecules and ions, in the Bénard and Zhabotinsky cases) are doing and the patterns they form are what they are *because* of their incorporation into the system-as-a-whole – in fact these are patterns *within* the systems in question. The parts would not be behaving as observed if they were not parts of that particular system (the 'whole'). The state of the system-as-a-whole is influencing (that is, acting like a 'cause' on) what the parts – the constituents – actually do. Many other examples of this kind could be taken from the literature on, for example, not only self-organizing and dissipative systems but also economic and social ones. Terrence Deacon (2001) has usefully categorized different kinds of emergent level.[8]

A wider use of 'causality' and 'causation' than Humean temporal, linear chains of causality as previously conceived (A→B→C...) is now needed to include the kind of whole–part, higher- to lower-level, relationships that the sciences have themselves recently been discovering in complex systems, especially the biological and neurological ones. One should perhaps better speak of 'determinative *influences*' rather than of 'causation', as having misleading connotations. Where such determinative influences of the whole of a system on its parts occurs, one is justified in attributing reality to those emergent properties and features of the whole system that have those consequences. Real entities have influence and play irreducible roles in adequate explanations of the world.

---

[8] Similar proposals are made by him in Deacon (2003). See also Weber and Deacon (2000).

Here the term 'whole–part influence' will usually be used to represent the net effect of all those ways in which a system-as-a-whole, operating from its 'higher' level, is a determining factor in what happens to its constituent parts at the 'lower' level. A holistic state, in this understanding, is determined by (is 'caused by', is 'a consequence of') a preceding holistic state *jointly* with the effects of its constituents with their individual properties in isolation. Such a 'joint' effect may be interpreted as a transmission of 'information' when this is conceived of in its broadest sense as that which influences patterns, forms of organization of constituents.[9] These ways in which interrelationships in complex systems in the natural world have been explicated provide clues, it will later be urged, to the concepts needed for relating the constituent entities of the God–world–humanity relation. But first it is necessary to examine two other significant developments, under pressure from scientific perceptions, in expounding the relation of God and the 'world', 'all-that-is', 'nature': namely, theistic naturalism and panentheism.

12.2 Theistic naturalism

The incessant pressure from the popular, widespread acceptance of the cogency of science has been towards

---

[9] J. C. Puddefoot has carefully clarified the relation between the different uses of 'information' (Puddefoot, 1991, pp. 7–25). First, physicists, communication engineers, and neuroscientists use it in referring to the probability of one outcome among many possible outcomes of a situation; second, there is the meaning of 'to give shape or form to' (stemming from the Latin *informare*); finally, the ordinary sense of information as knowledge, so, broadly, 'meaning'.

assuming, with little further consideration, that 'the world can best be accounted for by means of the categories of natural science (including biology and psychology) without recourse to the super-natural or transcendent as a means of explanation', which is a definition of 'naturalism'[10] or, according to the Oxford dictionary, 'A view of the world, and of man's relation to it, in which only the operation of natural (as opposed to supernatural or spiritual) laws and forces is assumed (1750)'.[11]

Such a stance precludes the divine. However, this is not a necessary consequence of taking account of the scientific knowledge of the world and its development, for a *theistic* naturalism may be expounded according to which natural processes, characterized by the laws and regularities discovered by the natural sciences, are themselves actions of God, who continuously gives them existence.

Contrary emphases have long historical roots since, for a century or more after Newton, creation still tended to be thought of as an act at a point in time when God created something external to God's self in a framework of an already existing space – not unlike the famous Michelangelo depiction of the creation of Adam on the ceiling of the Sistine Chapel. This led to a conception of God that was very 'deistic': God as external to nature, dwelling in an entirely different kind of 'space' and being of an entirely different 'substance' which by definition could not overlap or mix with that of the created order. In practice, and in spite of earlier theological insights, there

[10] See *A New Dictionary of Christian Theology* (Richardson and Bowden, 1983).
[11] See Shorter Oxford English Dictionary (1973).

was an excessive emphasis on God's transcendence and on the separation of God from what is created. However, cracks in this conceptual edifice began to appear in the late eighteenth century and early nineteenth century, when the age of the Earth inferred from geological studies was being stretched from 4004 BCE, deduced by adding up the ages of the Biblical patriarchs, to a process lasting many hundreds of thousands of years or more. But it was Darwin's eventually accepted proposal of a plausible mechanism for the changes in living organisms that led to the ultimate demise of the external, deistic notion of God's creative actions. In particular, those Anglican theologians who were recovering a sense of the sacramental character of the world stressed God's omnipresent creative activity in that world. Thus Aubrey Moore in 1889:

The one absolutely impossible conception of God, in the present day, is that which represents him as an occasional visitor. Science has pushed the deist's God further and further away, and at the moment when it seemed as if he would be thrust out all together Darwinism appeared, and, under the disguise of a foe, did the work of a friend. It has conferred upon philosophy and religion an inestimable benefit, by showing us that we must choose between two alternatives. Either God is everywhere present in nature, or He is nowhere.

(Moore, 1891, p. 73)

Moore and his co-religionists were not alone – the evangelical Presbyterian Henry Drummond saw God as working all the time through evolution:

Those who yield to the temptation to reserve a point here and there for special divine interposition are apt to forget that this virtually excludes God from the rest of the process. If God

326

appears periodically, He disappears periodically... Positively, the idea of an immanent God, which is the God of Evolution, is definitely grander than the occasional Wonder-worker, who is the God of an old theology.   (Drummond, 1894, p. 428)

Similarly Frederick Temple and also the Anglican Evangelical, Charles Kingsley, in *The Water Babies*, could affirm that 'God makes things make themselves' (Kingsley, 1930, p. 248).

For a theist, God must now be seen as acting to create in the world, often *through* what we call 'chance' operating within the created order, each stage of which constitutes the launching pad for the next. The Creator unfolds the created potentialities of the universe, in and through a process in which its possibilities and propensities become actualized. God may be said to have 'gifted' the universe, and goes on doing so, with a 'formational economy' that 'is sufficiently robust to make possible the actualization of all inanimate structures and all life forms that have ever appeared in the course of time' (van Till, 1998, pp. 349, 351).

So a revived emphasis on the *immanence* of God as Creator 'in, with and under' the natural processes of the world unveiled by the sciences becomes imperative if theology is to be in accord with all that the sciences have revealed since those debates of the nineteenth century. For a notable aspect of the scientific account on the natural world in general is the seamless character of the web that has been spun on the loom of time – at no point do modern natural scientists have to invoke any non-natural causes to explain their observations and inferences about the past. The processes that have occurred can, as we have just seen, be characterized as displaying *emergence*, for new forms of matter, and a hierarchy of organization of these forms themselves, appear in the course of time. New kinds of reality 'emerge' in time.

Hence the scientific perspective of the world, especially the living world, inexorably impresses upon us a *dynamic* picture of the world of entities, structures and processes involved in continuous and incessant change and in process without ceasing. This has impelled us to reintroduce into our understanding of God's creative relation to the world a dynamic element. This was always implicit in the Hebrew conception of a 'living God', dynamic in action, but has been obscured by the tendency to think of 'creation' as an event in the past. God has again to be conceived of as continuously creating, continuously giving existence to, what is new. God is creating at every moment of the world's existence in and through the perpetually endowed creativity of the very stuff of the world.

All of this reinforces the need to reaffirm more strongly than at any other time in the Christian (and Jewish and Islamic) traditions that in a very strong sense God is the immanent Creator, creating in and through the processes of the natural order. The processes are not themselves God, but the *action* of God-as-Creator. God gives existence in divinely created time to a process that itself brings forth the new – thereby God is creat*ing*. This means we do not have to look for any *extra* supposed gaps in which, or mechanisms whereby, God might be supposed to be acting as Creator in the living world.

A musical analogy might help: when we are listening to a musical work – say, a Beethoven piano sonata – then there are times when we are so deeply absorbed in it that for the

moment we are thinking Beethoven's musical thoughts
with him. If, however, anyone were to ask at that moment
(unseemingly interrupting our concentration!), 'Where is
Beethoven now?', we would have to reply that Beethoven-
as-composer was to be found only in the music itself.
Beethoven-as-composer is/was other than the music (he
'transcends' it), but his interaction with and communication to us is entirely subsumed in and represented by the
music itself – he is immanent in it and we need not look
elsewhere to meet him in that creative role. The processes
revealed by the sciences are in themselves God acting as
Creator, and God is not to be found as some kind of *additional* influence or factor added on to the processes of the
world God is creating. This perspective can properly be
called a *theistic naturalism*.

12.3 Panentheism

The scientific picture of the world points to a perspective
on God's relation to all natural events, entities, structures,
and processes in which they are continuously being given
existence by God, who thereby expresses in and through
them God's own inherent rationality. In principle this
should have raised no new problems for Western classical theism when it maintains the ontological distinction
between God and the created world. However, it often
conceived of God as a necessary 'substance' with attributes
and with a space 'outside' God in which the realm of the
created was, as it were, located. Furthermore, one entity
cannot exist *in* another and retain its own (ontological)
identity if they are regarded as substances. Hence, if God
is so regarded, God can only exert influence 'from outside'

on events in the world. Such 'intervention', for that is what it would be, raised acute problems in the light of our contemporary scientific perception of the causal nexus of the world being a closed one. Because of such considerations, this substantival way of speaking has become inadequate in my view and that of many others. It has become increasingly difficult to express the way in which God is present to the world in terms of 'substances', which by definition cannot be internally present to each other. This inadequacy of Western classical theism is aggravated by the evolutionary perspective which, as we have just seen, requires that natural processes in the world need to be regarded *as such* as God's creative action.

We therefore need a new model for expressing the closeness of God's presence to finite, natural events, entities, structures, and processes and we need it to be as close as possible without dissolving the distinction between Creator and what is created. In response to such considerations and those broad developments in the sciences, already outlined above, which engender a theistic naturalism, there has indeed been a 'quiet revolution' (Brierley, 2004, pp. 1–15)[12] in twentieth-century and early twenty-first-century theology by the resuscitation of pan*en*theism – the admittedly inelegant term for the belief that the Being of God includes and penetrates all-that-is, so that every part of it exists in God and (as against pantheism) that God's Being is more than, and is not exhausted by, it. In contrast to classical philosophical theism, with its reliance on the concept of necessary substance,

---

[12] The whole volume constitutes a survey of the various understandings, and misunderstandings, of the concept of panentheism.

panentheism takes embodied personhood for its model of God and so has a much stronger stress on the immanence of God in, with, and under the events of the world.

To say, as in the definition of panentheism, that the world is 'in' God evokes a spatial model of the God–world relation, as in Augustine's picture of the world as a sponge floating on the infinite sea of God (Augustine, 2006, VII, 7). This 'in' metaphor has advantages in this context over the 'separate-but-present-to' terminology of divine immanence in Western classical theism. For God is best conceived of as the circumambient Reality enclosing all existing entities, structures, and processes; and as operating in and through all, while being more than all. Hence, all that is not God has its existence within God's operation and Being. The infinity of God includes all other finite entities, structures, and processes – God's infinity comprehends and incorporates *all*. In this model, there is no 'place outside' the infinite God in which what is created could exist. God creates all-that-is *within* Godself.

One pointer to the cogency of a panentheistic interpretation of God's relation to the world is the way the different sciences interrelate to each other and to the world they study. A hierarchy of sciences from particle physics to ecology and sociology is required to investigate and explicate the embedded hierarchies of natural systems – as we saw, a 'layered' physicalism is implied, for the more complex is constituted of the less complex and all interact and interrelate in systems of systems. It is to this world so discovered by the sciences that we have to think of God as relating. The 'external' God of Western classical theism can be modelled as acting on such a world only by intervening separately at the various discrete levels. But if

God incorporates both the individual systems and the total System-of-systems within Godself, as in the panentheistic model, then it is readily conceivable that God could interact with all the complex systems at their own holistic levels. For God is present within the wholes as such, as well as to the parts.

At the terminus of one of the branching lines of natural hierarchies of complexity stands the human person – the complex of the human-brain-in-the-human-body-in-society. Persons can have intentions and purposes, which can be implemented by particular bodily actions. Indeed, the action of the body just *is* the intended action of the person. The physical action is describable, at the bodily level, in terms of the appropriate physiology, anatomy, etc., but it is also an expression of the intentions and purposes of the person's thinking. They are two modalities of the same psychosomatic event. To be embodied is a necessary condition for persons to have perception, to exert agency, to be free, and to participate in community. Personal agency has been used both traditionally in the Biblical literature and in contemporary theology as being an appropriate model for God's action in the world. Our intentions and purposes seem to transcend our bodies, yet in fact are closely related to brain events and can be implemented in the world only through our bodies. Our bodies are indeed ourselves under one description and from another perspective. In personal agency there is an intimate and essential link between what we intend and what happens to our bodies. Yet 'we' as thinking, conscious persons appear to transcend our bodies while nevertheless being immanent in them. This 'psychosomatic', unified understanding of human personhood reinforces the use of a panentheistic model for God's relation to the world. For, according to that model, God is *internally* present to all of its entities, structures, and processes in a way that can be regarded as analogous to the way we as persons are present and act in our bodies. This model, in the light of current concepts of the person as a psychosomatic unity, is then an apt way of modelling God's *personal* agency in the world.

As with all analogies, models, and metaphors, qualifications are needed before we too readily draw a parallel between God's relation to the world and our relation as persons to our bodies. The first is that the God who, we are postulating, relates to the world like a personal agent, is also the one who creates it, gives it existence, and infinitely transcends it. Indeed, the panentheistic model emphasizes this in its 'more than the world'. Moreover *we* do not create our own bodies. The second qualification of the model is that as human persons, we are not conscious of most of what goes on in our bodies in autonomous functions such as breathing, digestion, heart beating. Yet other events in our bodies are conscious and deliberate, as we have just been considering. So we have to distinguish between these; but this can scarcely apply to an omniscient God's relation to the world. The third qualification of the model is that, in so using human personal agency as similar to the way God interacts with the world, we are not implying that God is 'a person' – rather that God is more coherently thought of as 'at least personal', indeed as 'more than personal' (recall the 'more than' of panentheism). Perhaps we could even say that God is 'suprapersonal' or 'trans-personal', for there are some essential aspects to God's nature that cannot be subsumed under the categories applicable to human persons. In my view,

333

the pantheistic model allows one to combine a renewed and stronger emphasis on the immanence of God in the world with God's ultimate transcendence over it. It does so in a way that makes the analogy of personal agency both more pertinent and less vulnerable than the Western, externalist model to the distortions corrected by the above qualifications of the model of the world-as-God's-body.

The fact of natural (as distinct from human, moral) evil continues to be a challenge to belief in a benevolent God. In the classical perception of God as transcendent and as existing in a 'space' distinct from that of the world, there is an implied detachment from the world in its suffering. This thereby renders the 'problem of evil' particularly acute. For God can do something about evil only by an intervention from outside, which provokes the classical dilemma of either God can and will not, or would but cannot: God is either not good or not omnipotent. But an ineliminable hard core of offence remains, especially when encountered directly, and often tragically, in personal experience. For the God of classical theism witnesses, but is not involved *in*, the sufferings of the world – even when closely 'present to' and 'alongside' them.

Hence, when faced with this ubiquity of pain, suffering, and death in the evolution of the living world, we are impelled to infer that God, to be anything like the God who *is* Love in Christian belief, must be understood to be suffering in, with, and under the creative processes of the world. Creation is costly *to God*, we concluded. Now, when the natural world, with all its suffering, is pantheistically conceived of as 'in God', it follows that the evils of pain, suffering, and death in the world are internal to God's own self. So God must have experience

334

of the natural. This intimate and actual experience of God must also include all those events that constitute the evil intentions of human beings and their implementation – that is, the 'moral evil' of the world of human society. God is creating the world from within and, the world being 'in' God, God experiences its sufferings directly as God's own and not from the outside.

12.4 The consequences for theology of the reappraisal of the natural

It is clear from the foregoing that the sciences have led during the last few decades to a fundamental reassessment of the nature and history of the world, which now, it has been argued above, has to be characterized in terms of an *emergentist monism*. The world is a hierarchy of inter-locking complex systems and it has come to be recognized that these complex systems have a determinative effect, an exercising therefore of causal powers, on their components – a whole–part influence. This in itself implied an attribution of reality to the complexes and to their properties that undermines any purely reductionist under-standing and suggested that the determinative power of complex systems on their components can often best be understood as a flow of 'information', understood in its most general sense as a pattern-forming influence. Suc-cessive states of a complex (and most natural entities *are* complexes) are what they are as a result of a *joint* effect of the state of the complexes-as-a-whole and of the proper-ties of the individual components. The wholes and parts are intimately interlocked regarding their properties, and so in the very existence that a creator God gives them.

335

Furthermore, the evolutionary perspective of the sciences, cosmological and biological, entails an understanding of creation by God as a continuous activity – dynamic models and metaphors of divine creation become necessary. The work of God as Creator is manifested all the time in the natural processes, which are revealed in all their regularities by the sciences. These processes have now to be seen as both natural and divine, so that a *theistic naturalism* has had to be affirmed according to which God is at work creating and maintaining the natural order and is not conceived of as external to it, occasionally intervening and disrupting its regularities. This involves a much tighter linkage between the divine and the natural, and so of the immanence of the divine, than has prevailed in much Western Christian theology.

These developments in the understanding of God's relation to the natural, albeit created, order, which have arisen from pressure from the new scientific perspectives on the world and its history, are integrated in the model of *panentheism*. This model also attempts to restore the necessary balance between God's transcendence and God's immanence, which, in this context, had become over-weighted in favour of the former because the relation between God and the world had too often been couched in terms of substances. The panentheistic model is, in fact, closer to an older tradition within theology than has usually been realized. For the Eastern Christian tradition has been explicitly panentheistic in holding together God's transcendence and immanence.[13] For example, Gregory Palamas (c.1296–1359 CE) made a distinction-in-unity

---

[13] According to Bishop Kallistos of Diokleia (1997, pp. 12–14).

between God's essence and his energies; and Maximus the Confessor (c.580–662 CE) regarded the Creator-*Logos* as being characteristically present in each created thing as God's intention for it, its inner essence (its *logoi*), which makes it distinctively itself and draws it towards God.

What is transmitted across the 'interface' between God and the world may perhaps best be conceived of as something like a flow of information – a pattern-forming influence. However, one has to admit that, because of the 'ontological gap(s)' between God and the world, which must always exist in any theistic model, this is only an attempt at making intelligible that which we can postulate as being the effect of God seen, as it were, from the human side of the boundary. Whether or not this use of the notion of information flow proves helpful in this context, we do need some way of indicating that the effect of God at all levels is that of pattern-shaping in its most general sense. I am encouraged in this kind of exploration by the recognition that the Johannine concept of the *Logos*, the Word, of God may be taken to emphasize God's creative patterning of the world and so as God's self-expression *in* the world.

An overwhelming impression is given by these developments, in both the philosophy of science (emergentist monism) and in theology (theistic naturalism and panentheism), of the world as an interlocking System-of-systems saturated, as it were, with the presence of God shaping patterns at all levels. This enhanced emphasis on divine immanence in natural events warrants, I suggest, the application of the same interpretative concepts used for them in those complex events {God + nature + persons} that constitute the human experience of God and that are the usual focus of theological discourse. In the following, an

337

attempt is made to employ theological language about the relevant relationships and properties in various contexts in an emergentist monist manner, as in the interpretation of natural systems – themselves now conceived of as exemplifying the activity of God, according to theistic naturalism, and as being 'in' God, according to panentheism. For brevity, we will, in the following, denote this fusion of these horizons as *EPN* (= *emergentist/monist–panentheistic-naturalist*).

Hence, in examining religious experiences requiring distinctive theological interpretation – putative 'complexes' of {God + nature + persons} – we might hope to find, if only partially, aspects that are: *ontological* with respect to the emergence of new kinds of reality requiring distinctive language and concepts and referring to *what is there*; and *causal*[14] with respect to the description of whole–part *influences* of 'higher' levels (God, in these instances) on the 'lower' ones and referring to *what is being effected and transmitted* in it (and thereby paralleling the pattern-forming associated with the transmission of information in natural systems).

We examine now some of those areas of discourse concerning theological themes that might be illuminated by the recognition that belief in God as Creator involves the recognition, which I have been describing, of the character of the processes whereby God actually creates new forms, new entities, structures, and processes. They emerge with new capabilities, requiring distinctive language on our part

---

[14] It cannot be too strongly emphasized that 'causal' does not here, and elsewhere in this chapter, refer to a Humean succession of cause/effect events, but to the influence of the state of the whole of a complex on the properties and behaviour of its constituent parts.

to distinguish them. If God is now recognized as present in, with, and under this whole process, then our theology should be able to recognize those same ontological and causal-influential features that are expressed in the concepts of emergent monism, theistic naturalism, and panentheism (the whole *EPN* perspective) as also evident in these other modes of God's presence.

12.4.1 Incarnation

Nearly three decades have elapsed since the publication of *The Myth of God Incarnate* (Hick, 1977) and the consequent flurry of books it evinced. A range of 'Christologies' continue to represent the beliefs of Christians of various colours, with the more orthodox appealing to the Definition of Chalcedon of 451 CE as the basis of the formulation of their faith, while others, less concerned for conformity, seek other more dynamic modes of expression to describe how they believe 'God was in Christ reconciling the world to himself'.[15] It is, at least, widely recognized that that famous Definition, with its uncompromising 'two natures' in 'one person', afforded only boundaries within which Christian discourse was urged to range. The paradox implicit in this assertion has continued to be a theological gadfly – an unavoidable assertion, apparently, but with no satisfying resolution of its paradox.

The recognition of the all-pervasiveness of emergence as a feature of the world provides now, I would suggest, a way of drawing the sting of this paradox. For complex systems, as we saw, themselves display properties that are

---

[15] See 2 Corinthians 5, 19.

the *joint* outcomes of those of its components and of the system-as-a-whole. This allows one to refer to the new reality constituted by the whole without any contradiction of the different reality of the components. Applying this way of thinking to the person of the historical Jesus leads to the following proposal. The creation of the human personhood of Jesus, born of Mary and fully human, is in this *EPN* perspective in itself a divine action[16] at the biological level, corresponding to the component element of a system. At the level of the historical Jesus – the macro level, as it were – Jesus' will was fully open in submission to the divine will, so that in his total person there emerged a unique manifestation of the divine being insofar as that is expressible in human form – the manifestation of self-offering love. There is, in this *EPN* perspective, no basic contradiction between the human level of Jesus' existence ('born of Mary, etc.'), itself a divine expression, and the expression of the divine in the total personhood of Jesus, those features of the encounter with him that led his followers to perceive a dimension of transcendence characteristic only of the divine. The relation between the two is that of emergence: the way God creates new entities in the natural world, it has transpired.

Jesus' Jewish followers encountered in him (especially in his Resurrection) a dimension of divine transcendence that, as devout monotheists, they had attributed to God alone. But they also encountered him as a complete human being, and so experienced an intensity of God's immanence in the world different from anything else in their experience or tradition. Thus it was that the fusion of these

---

[16] Whether or not one affirms the virginal conception.

two aspects of their awareness – that it was *God* acting in and through Jesus the Christ – gave rise to the conviction that something new had appeared in the world of immense significance for humanity. A new emergent had appeared within created humanity. Thus it was, too, that they ransacked their cultural stock of available images and models (for example 'Christ' = 'Messiah' (= 'Anointed'), 'son of God', 'Lord', 'Wisdom', '*Logos*'), at first Hebraic and later Hellenistic, to give expression to this new, nonreducible, distinctive mode of being and becoming, instantiated in Jesus the Christ. One might say that God 'informs' the human personhood of Jesus such that God's self-expression occurs in and through Jesus' humanity.

When we reflect on the significance of what the early witnesses reported as their experience of Jesus the Christ, we find ourselves implicitly emphasizing both the *continuity* of Jesus with the rest of humanity, and so with the rest of nature within which *Homo sapiens* evolved, and, at the same time, the apparent *discontinuity* constituted by what is distinctive in his relation to God and what, through him (his teaching, life, death, and Resurrection), the early witnesses experienced of God. This combination of continuity with discontinuity is just what we have come to recognize in the emergent character of the natural world, and it seems appropriate to apply this to the cluster of notions concerning the person of Jesus. In Jesus the Christ a new reality has emerged and a new *ontology* is necessitated – hence the classical imagery just cited.

How then, in the light of this, might we interpret the experience of God that was mediated to his disciples and to the New Testament church through Jesus? That is, how can we understand the Christ-event as God's self-communication and interaction with the world such that it is intelligible in the light of today's natural and human sciences? We need to explicate in these terms the conclusions of scholars about the understanding in New Testament times of Jesus the Christ. Note, for example, J. G. Dunn's conclusion that: 'Initially Christ was thought of...as the climactic embodiment of God's power and purpose God himself reaching out to men...God's creative wisdom...God's revelatory word...God's clearest self-expression, God's last word' (Dunn, 1980, p. 262, emphasis omitted).

These descriptions of what Jesus the Christ was to those who encountered him and the early church are all, in their various ways, about God *communicating* to humanity. In the broad sense in which we have been using the terms, they are about an 'input of information'. This process of 'input of information' from God conforms with the actual content of human experience, as the conveying of 'meaning' from God to humanity.[17] God can convey divine meanings through events and patterns of events in the created world – those in question here are the life, teaching, death, and Resurrection of the human person, Jesus of Nazareth, as reported by these early witnesses. As the investigations of the New Testament show, they experienced in Jesus, in his very person and personal history, a communication from God, a revelation of God's meanings for humanity. So it is no wonder that, in the later stages of reflection in the New Testament period, John conflated the concept of divine Wisdom with that of the *Logos*, the 'Word' of God, in order to say what he

---

[17] A theme developed in Peacocke (1996, pp. 321–339).

intended about the meaning of Jesus the Christ for the early witnesses and their immediate successors. The *locus classicus* of this exposition is, of course, the prologue to the gospel of John. John Macquarrie (1990, pp. 43–44, 106–108) notes that the expression 'Word' or *Logos*, when applied to Jesus, not only carries undertones of the image of 'Wisdom', it also conflates two other concepts: the Hebrew idea of the 'word of the Lord' for the will of God expressed in utterance, especially to the prophets, and in creative activity; and that of '*logos*' in Hellenistic Judaism, especially in Philo – the Divine Logos, the creative principle of rationality operative in the universe, especially manifest in human reason, formed within the mind of God, and projected into objectivity. He suggests substituting 'Meaning' for Word–*Logos*, as it helps to convey better the Gospel's affirmation of what happened in creation and in Jesus the Christ, since the conveying of meaning, in the ordinary sense, is implemented initially by an input of 'information' – that new reality, the Incarnation, involves distinctive causal influences.

The ideas that generated the *EPN* perspective do indeed seem to illuminate what Christians wish today to affirm about Jesus the Christ as a unique revelation from God about humanity and about God's own self.

12.4.2 The Eucharist

The relations of humanity to God may also be illuminated by our understanding of the emergence of new realities in complex, especially self-organizing, systems. For in many situations where God is experienced by human persons we have by intention and according to well-winnowed

343

experience and tradition complexes of interacting personal entities, material things, and historical circumstances that are not reducible to concepts applicable to these individual components. Could not new realities and so new experiences of God for humanity be seen to 'emerge' in such complexes and even to be causally effective?

I am thinking,[18] for example, of the Church's Eucharist (Holy Communion, the Mass, 'The Lord's Supper') in which there exists a distinctive complex of interrelations between its constituents. These latter could be identified inter alia (for it is many-layered in the richness of its meanings and symbols) as follows.

(1) Individual Christians are motivated by a sense of *obedience* to the ancient, well-authenticated historically, command of Jesus, the Founder of their faith, at the actual Last Supper to 'Do *this...*', that is to eat the bread and to drink the wine in the same way he did on that occasion and so to identify themselves with his project in the world.

(2) Christians of all denominations have been concerned that their communal act is properly *authorized* as being in continuity with that original act of Jesus and its repetition, recorded in the New Testament, in the first community of Christians. Churches have differed about the character of this authorization but not about its importance.

(3) The physical 'elements', as they are often called, of bread and wine are, of course, part of the matter of

---

[18] An interpretation of the Eucharist was originally suggested in Peacocke (1972, pp. 28–37, especially p. 32); and (with some additions) in Peacocke (1994, pp. 124–125). It is entirely congruent with that recently expounded by N. H. Gregersen (2000, pp. 180–182).

the world and so representative, in this regard, of the created order. So Christians perceive in these actions, in this context and with the words of Jesus in mind, that a *new significance and valuation of the very stuff of the world* is being expressed in this action.

(4) Because it is bread, and not corn, wine, and not grapes, that are consecrated, this act has come to be experienced also as a new evaluation of the work of *humanity in co-creating with God in ordinary work.*

(5) The broken bread and poured-out wine was explicitly linked by Jesus with his anticipated self-sacrificial offering of himself on the cross in which his body was broken and blood shed to draw all towards unity of human life with God. Christians in this act consciously acknowledge and identify themselves with Jesus' self-sacrifice, thereby offering to reproduce the same *self-emptying love* for others in their own lives and so to further his purposes of bringing in the Reign of God in the world.

(6) They are also aware of the promise of Jesus to be present again in their re-calling and re-making of the historical events of his death and resurrection. This 'making present' (*anamnesis*) of the Jesus who is regarded as now fully in the presence of – and is, in some sense, identified with – God is a unique and spiritually powerful feature of this communal act.

(7) The creative *presence of God*, as transcendent, incarnate, and immanent.

Here, do we not have an exemplification of the emergence of a new kind of reality requiring a distinctive ontology? For what (if one dare so put it) 'emerges' in the eucharistic event *in toto* can only be described in special nonreducible terms such as 'Real Presence'. Moreover, a distinctive kind of divine, transformative causality is operative in the sacramental experience of the participants. Hence a distinctive terminology of causal influences is also necessitated (involving terms such as 'Sacrifice'), for in the sacrament there is an effect on both the individual and on the community that induces distinctively Christian personhood and society (of 'being ever deeper incorporated into this body of love').[19] So it is not surprising that there is a branch of study called 'sacramental theology' to explicate this special reality and human experience and interpretations of it. Since God is present 'in, with, and under' this holistic eucharistic event, in it God may properly be regarded as through it distinctively acting on the individual and community – surely an exemplification of God's non-intervening, but specific, 'whole–part' influence on the world in the *EPN* perspective.

12.4.3 God's interaction with the world

In a world that is a closed causal nexus, increasingly explicated by the sciences, how might God be conceived of as influencing particular events, or patterns of events, in the world without interrupting the regularities observed at the various levels studied by the sciences? A model I have proposed[20] is based on the recognition that the omniscient God uniquely knows, over all frameworks of reference of

---

[19] See Gregersen (2000).
[20] For an elaboration of this move, see Peacocke (1993, pp. 160–166). For the history and development of this proposal, see Peacocke (1995, note 1, p. 263); and also Peacocke (1999, *note* 1, p. 215).

time and space, everything that it is possible to know about the state(s) of all-that-is, including the interconnectedness and interdependence of the world's entities, structures, and processes. This is a pan*en*theistic perspective, for it conceives of the world as, in some sense, being 'in' God, who is also 'more' than the world. It also follows that the world would be subject to any divine determinative influences that do not involve matter or energy (or forces). Thus, mediated by such whole–part influences on the world-as-a-whole (as a *System*-of-systems) and thereby on its constituents, God could cause particular events and patterns of events, which express God's intentions, to occur. These would then be the result of 'special, divine action', as distinct from the divine holding in existence of all-that-is, and so would not otherwise have happened had God not so intended. By analogy with the exercise of whole–part influence in the natural systems already discussed, such a unitive, holistic effect of God on the world could occur without abrogating[21] any of the laws (regularities) that apply to the levels of the world's constituents. This influence would be distinguished from God's universal creative action, in that particular intentions of God for particular patterns of events to occur are thereby effected; inter alia, patterns of events could be intended by God in response to human actions or prayers.

The ontological 'interface' at which God must be deemed to be influencing the world is, on this model, that

---

[21] The same may be said of *human* agency in the world. Note also that this proposal recognizes more explicitly than is usually expressed that the 'laws' and regularities that constitute the sciences usually apply only to certain perceived, if ill-defined, levels within the complex hierarchies of nature.

which occurs between God and the totality of the world (= all-that-is), and this may be conceived of panentheistically as within God's own self. What passes across this 'interface', I have also suggested (Peacocke, 1993, pp. 161, 164),[22] may perhaps be conceived of as something like a flow of information – a pattern-forming influence. Of course, one has to admit that, because of the 'ontological gap(s)' between God and the world that must always exist in any theistic model, this is only an attempt at making intelligible that which we can postulate as being the initial effect of God experienced from, as it were, our side of the ontological boundary.[23] Whether or not this use of the notion of information flow proves helpful in this context, we do need some way of indicating that the effect of God at this, and hence at all levels, is that of pattern-shaping in its most general sense. I am again encouraged in this kind of exploration by the recognition that the Johannine concept of the *Logos*, the Word of God, may be taken to emphasize God's creative patterning of the world and also God's self-expression *in* the world.

On this model, the question arises at what level or levels in the world such divine influences might be coherently conceived as acting. By analogy with the operation of whole–part influence in natural systems, I have in the past suggested that, because the 'ontological gap(s)' between the world and God is/are located simply *everywhere* in space and time, God could affect holistically the state of the

---

[22] J. Polkinghorne has made a similar proposal in terms of the divine input of 'active information' (Polkinghorne, 1996, p. 36–37).

[23] I would not wish to tie the proposed model too tightly to a 'flow of information' interpretation of the mind–brain–body problem.

world (the whole in this context) at all levels. Understood in this way, the proposal implies that patterns of events at the physical, biological, human, and even social levels could be influenced by divine intention without abrogating natural regularities at any of these levels. In this form it poses in a particularly acute form the challenge of 'special divine action' to current scientific understandings of the world as a closed nexus of webs of causes and whole–part influences. The sharpness of this challenge is arguably less if the top-down influence of God is conceived as operating mainly, or even exclusively, at the level of the human person, the emergent reality of which can be located at the apex of the systems-based complexities of the world. God would then be thought of as acting in the world in a whole–part, top-down manner by shaping human personal experience, which thereby effects events at the physical, biological, and social levels.

These two limiting forms of the proposal of special divine action by top-down divine influence are not mutually exclusive. However, divine action in a form that is confined to the personal level is less challenged by (has more 'traction' with) the general scientific account of the world than when such divine action is proposed to be at *all* levels. At this stage in my formulating this proposal, I am inclined to postulate divine top-down influences at all levels, but with an increasing intensity and precision of location in time from the lowest physical levels up to the personal level, where it could be at its most intense and most focused. More general theological considerations need to be brought to bear on how to formulate this model of special divine action. One relevant consideration might be developed as follows.

I hope the model as described so far has a degree of plausibility in that it depends only on an analogy with complex natural systems in general and on the way whole–part influence operates in them. It is, however, clearly too impersonal to do justice to the *personal* character of many (but not all) of the profoundest human experiences of God. So there is little doubt that it needs to be rendered more cogent by recognizing, as I have argued above, that among natural systems the instance par excellence of whole–part influence in a complex system is that of personal agency. Indeed, I could not avoid speaking above of God's 'intentions' and implying that, like human persons, God had purposes to be implemented in the world. For if God is going to affect events and patterns of events in the world, we cannot avoid attributing personal predicates such as intentions and purposes to God – inadequate and easily misunderstood as they are. So we have to say that, although God is ineffable and ultimately unknowable in essence, yet God 'is at least personal', and personal language attributed to God is less misleading than saying nothing! That being so, we can now legitimately turn to the exemplification of whole–part influence in the mind–brain–body relation as a resource for modelling God's interaction with the world. When we do so, the cogency of the 'personal' as a category for explicating the wholeness of human agency reasserts itself and the traditional, indeed Biblical, model of God as in some sense a 'personal' agent in the world, acting especially on persons, is rehabilitated – but now in that quite different metaphysical, non-dualist framework of the *EPN* perspective, which is itself coherent with the world view that the sciences engender.

12.5 Conclusion

I propose that the principles involved in trying to make clear what is special about these and other spiritual situations involving {God + nature + persons} is broadly applicable[24] to many other experiences of theological concern and interest, both historical and contemporary. Thus 'grace' may be conceived as a causally influential and transformative effect of God on a human being that operates when a person comes into an intimate relation with God under the particular circumstances that characterize its different forms.[25] Transformation is a key feature of the effects of grace, so there is always an ontological aspect to the experience as well as a causal-influential one. These principles are also applicable to the understanding of intercessory prayer in which the 'complex' under consideration involves more than one person, although one in particular is the object of the prayer(s). The whole complex of {God + persons} can be conceived of as constituting a new kind of reality with new causal-influential capacities that are *sui generis*.

The *EPN* perspective I have been trying to expound, in conjunction with the new sciences of complexity and of self-organization, provide, it seems, a fruitful and illuminating release for theology from the oppression of excessively reductionist interpretations of the hierarchy of the

---

[24] A plea I have made elsewhere (Peacocke, 2003, pp. 201–202), in an approach I have long since adumbrated in my Bampton lectures of 1978 (Peacocke, 1979, pp. 367–371).
[25] Grace has been variously, and somewhat over-formally, classified as inter alia 'extrinsic', 'uncreated', 'created', 'habitual', 'sanctifying', 'actual', 'elevating', 'prevenient', 'efficacious', 'sufficient', etc.

sciences, and a making accessible of theological language and concepts to the general exchanges of the intellectual life of our times, a milieu from which it has been woefully and misguidedly excluded for too long.

The new insights into the complexifying and information-bearing capacity of matter have generated a metaphysic in emergentist monism that, as interpreted in a panentheistic and natural-theistic framework, allows one to see a congruence and contiguity between the nature of matter and the experiences that theology seeks to articulate. Hence, in responding to one of the questions this volume poses – What is ultimate? – one does, now, not have to choose *between* 'God, matter, and information' but can hold them altogether in a new kind of synthesis that obviates the false dichotomies of the sciences/humanities, matter/spirit, and science/religion that have plagued Western culture for too long.

References

Augustine, S. (2006). *Confessions*, trans F. W. P. Brown and F. J. Sheed. Indianapolis, IN: Hackett Publishing Company.

Bechtel, W., and Abrahamson, A. (1991). *Connectionism and the Mind*. Cambridge, MA: Blackwell Publishers.

Bechtel, W., and Richardson, R. C. (1992). Emergent phenomena and complex systems. In *Emergence or Reduction? Essays on the Prospects of Nonreductive Physicalism*, eds A. Beckermann, H. Flohr, and J. Kim. Berlin: Walter de Gruyter Verlag, 257–288.

Bishop Kallistos of Diokleia (1997). Through the creation to the creator. *Ecotheology*, 2: 8–30.

Brierley, M. W. (2004). Naming a quiet revolution: The panentheistic turn in modern theology. In *In Whom We Live and*

Move and Have Our Being: Panentheistic Reflections on God's Presence in a Scientific World, eds P. Clayton and A. Peacocke. Grand Rapids, Michigan: Eerdmans, 1–15.

Campbell, D. T. (1974). 'Downward causation' in hierarchically organised systems. In *Studies in the Philosophy of Biology: Reduction and Related Problems*, eds F. J. Ayala and T. Dobhzhansky. London: Macmillan, 179–186.

Deacon, T. (2001). Three Levels of Emergent Phenomena. Paper presented to the Science and Spiritual Quest Boston Conference, October 21–23.

Deacon, T. (2003). The hierarchic logic of emergence: Untangling the interdependence of evolution and self-organization. In *Evolution and Learning: The Baldwin Effect Reconsidered*, eds B. Weber and D. Depew. Cambridge, MA: MIT Press, 273–308.

Drummond, H. (1894). *The Lowell Lectures on the Ascent of Man*. London: Hodder & Stoughton.

Dunn, J. G. (1980). *Christology in the Making*. London: SCM Press.

Gregersen, N. H. (2000). God's public traffic: Holist versus physicalist supervenience. In *The Human Person and Theology*, eds N. H. Gregersen, W. B. Drees, and U. Görman. Edinburgh: T & T Clark, 153–188.

Hick, J. (1977). *The Myth of God Incarnate*. London: SCM Press.

Kim, J. (1992). 'Downward causation' in emergentism and nonreductive physicalism. In *Emergence or Reduction? Essays on the Prospects of Nonreductive Physicalism*, eds A. Beckermann, H. Flohr, and J. Kim. New York: Walter de Gruyter, 134–135.

Kim, J. (1993). Non-reductivism and mental causation. In *Mental Causation*, eds J. Heil and A. Mele. Oxford: Clarendon Press, 204.

Kingsley, C. (1930). *The Water Babies*. London: Hodder & Stoughton.

Macquarrie, J. (1990). *Jesus Christ in Modern Thought*. London: SCM Press.

Moore, A. (1891). The Christian doctrine of God. In *Lux Mundi*, 12th ed, ed. C. Gore. London: Murray, 41–81.

Morowitz, H. (2002). *The Emergence of Everything: How the World Became Complex*. New York: Oxford University Press.

Peacocke, A. R. (1972). Matter in the theological and scientific perspectives – a sacramental view. In *Thinking about the Eucharist. Papers by members of the Church of England Doctrine Commission*, ed. I. T. Ramsey. London: SCM Press, 28–37.

Peacocke, A. R. (1979). *Creation and the World of Science*. Bampton lectures, Oxford: Clarendon Press.

Peacocke, A. R. (1983/1989). *The Physical Chemistry of Biological Organization*. Oxford: Clarendon Press.

Peacocke, A. R. (1993). *Theology for a Scientific Age: Being and Becoming – Natural, Divine and Human*, 2nd enlarged ed. Minneapolis: Fortress Press, and London: SCM Press.

Peacocke, A. R. (1994). *God and the New Biology*. Gloucester: Peter Smith.

Peacocke, A. R. (1995). God's interaction with the world: The implications of deterministic 'chaos' and of interconnected and interdependent complexity. In *Chaos and Complexity: Scientific Perspectives on Divine Action*, eds R. J. Russell, N. Murphy, and A. R. Peacocke. Berkeley: Vatican Observatory Publications, Vatican City State, and the Center for Theology and the Natural Sciences, 263–287.

Peacocke, A. R. (1996). The incarnation of the informing self-expressive word of God. In *Religion and Science: History, Method, Dialogue*, eds W. M. Richardson and W. J. Wildman. New York and London: Routledge, 321–339.

Peacocke, A. R. (1999). The sound of sheer silence: How does God communicate with humanity? In *Neuroscience and the Person: Scientific Perspectives on Divine Action*, eds R. J.

Russell et al. Berkeley: Vatican Observatory Publications, Vatican City State and the Center for Theology and the Natural Sciences, 215–247.

Peacocke, A. R. (2001). *Paths from Science towards God – the End of All Our Exploring*. Oxford: Oneworld.

Peacocke, A. R. (2003). Complexity, emergence and divine creativity. In *From Complexity to Life*, ed. N. H. Gregersen. Oxford: Oxford University Press, 187–205.

Peacocke, A. R. (2007). Emergent realities with causal efficacy – some philosophical and theological applications. In *Evolution and Emergence: Systems, Organisms, Persons*, eds N. Murphy and W. R. Stoeger. Oxford: Oxford University Press, 267–283.

Polkinghorne, J. C. (1996). *Scientists as Theologians*. London: SPCK.

Prigogine I., and Stengers, I. (1984). *Order Out of Chaos*. London: Heinemann.

Puddefoot, J. C. (1991). Information and creation. In *The Science and Theology of Information*, eds C. Wassermann, R. Kirby, and B. Rordoff. University of Geneva: Editions Labor et Fides.

Richardson, A., and Bowden, J., eds. (1983). *A New Dictionary of Christian Theology*. London: SCM Press.

Richardson, R. C. (1992). Emergent phenomena and complex systems. In *Emergence or Reduction? Essays on the Prospects of Nonreductive Physicalism*, eds A. Beckermann, H. Flohr, and J. Kim. Berlin and New York: de Gruyter, 257–288.

Shorter Oxford English Dictionary (1973). Oxford: Oxford University Press.

van Till, H. (1998). The creation: Intelligently designed or optimally equipped. *Theology Today*, 55: 344–364.

Weber, B., and Deacon, T. (2000). Thermodynamic cycles, developmental systems, and emergence. *Cybernetics & Human Knowing*, 7: 21–43.

Wimsatt, W. C. (1981). Robustness, reliability and multiple-determination in science. In *Knowing and Validating in the Social Sciences: A Tribute to Donald T. Campbell*, eds M. Brewer and B. Collins. San Francisco: Jossey-Bass, 124–163.

13

God as the ultimate
informational principle

KEITH WARD

∼

Scientists who speculate on philosophical questions usually agree that classical materialism – the view that reality consists of nothing but small massy particles bumping
into one another in an absolute and unique space–time –
is intellectually dead. Accounts of the universe now regularly involve notions such as that of manifold space–times,
quantum realities that exist at a more ultimate level than,
and are very different from, massy particles in one specific
space, and informational codes that contain instructions
for building complex integrated structures displaying new
sorts of emergent property.

What this suggests is that the nature of the reality investigated by physics and biology is much more complex and
mysterious than some Newtonian materialists thought
(though of course Newton himself was as far from being a
materialist as one can get). In particular, the role of information in any account of our universe has come to take
on a new importance.

Information and the Nature of Reality: From Physics to Metaphysics, eds. Paul Davies
and Niels Henrik Gregersen. Published by Cambridge University Press
© P. Davies and N. Gregersen 2010, 2014.

Most contributors to this volume distinguish three main types of information – Shannon information, "shaping" information, and semantic information.[1]

Shannon information is a matter of how to input the maximum amount of information into a closed physical system. It is concerned, it might be said, with quantity rather than quality, in that it totally ignores questions of the significance or function of the information that a physical system might contain. This is a technical matter for information technologists, and I shall not consider it further.

The second is "shaping" or "coding" information: the sort of thing we might have in mind when thinking of how DNA carries the information for constructing proteins and organic cells and bodies. We can understand what DNA is only when we see not only its chemical composition, but also how that composition leads to the construction of bodies.

Few biologists, however, think that this function of DNA is actually designed, in the sense of being intentionally set up in order to achieve the purpose of building a body. DNA, it is very widely thought, has evolved by processes of random mutation and natural selection to be an efficient replicating machine, which uses bodies as an aid to replication, but has come to do so by entirely blind and randomly evolved means.

For this view, the use of functional language is perhaps necessary as a shortcut for understanding the more basic chemical processes, which are far too complex to be spelled out in detail. This is epistemological emergence

[1] See Puddefoot (1991, pp. 7–25).

with a reductionist ontology – the basic mechanisms are all ordinary chemical ones, but it is easier for us to understand them if we speak of functions and codes that can be "read" and interpreted by ribosomes. We could reduce this language to that of chemistry, but it is too cumbersome to bother.

A rather different view is that we could not even in principle reduce the language of biology or psychology to that of chemistry or physics. Even though no new physical entities are involved, the way the basic physical entities interrelate and organize means that integrated and complex entities act in accordance with new principles, not deductively derivable from nor reducible to those of their simpler physical constituents.

So, for instance, the laws of nations are not reducible to laws governing the relation of all their constituent persons, but nations contain no entities but persons. It is their organization into complex structures that produces new principles of interaction, though it produces no new physical entities (nations are not super-persons). Such new principles of interaction might be informational, in that some parts provide the information that governs the behavior of other parts within the whole, or that enables the whole to be constructed as a complex entity.

It seems as though the position of an entity within a structure, and the forms of its relation to other entities in that structure, call forth new principles of interaction, causing it to function as a part of a complex integrated totality.

New laws of nature, new ways of interaction, emerge that are not just reducible to the laws of interacting particles considered in isolation. Structure becomes important to understanding. Many informational systems may be understood as having a specific function within an integrated totality that emerges only when that totality exists as a system.

These facts have led some scientists to speak of holistic explanation – explanation of elementary parts in terms of a greater whole – as an appropriate form of scientific explanation. Some, especially quantum physicists, extend the idea of holistic explanation to the whole universe, considered as a total physical system.

13.1 A mathematical possibility space

Recent hypotheses in quantum physics suggest that the whole physical universe is "entangled" in such a way that the parts of a system – even the behavior of elementary particles – cannot be fully understood without seeing their role within a greater whole: ultimately the whole of space–time. There may be no non-physical bits of "stuff," but there seem to be laws of their interaction that can be specified only from a grasp of whole systems, rather than atomistically. In quantum cosmology we are encouraged to see the whole universe as a complex system, and to think that knowledge of the total system may be needed fully to explain the behavior of its simple parts.[2]

Perhaps the origin of the universe, the explanation of which is the elusive Holy Grail of cosmology, can be fully understood only when its fullest development is understood, and we see its simplest and earliest parts as necessarily implied by the fully developed structure in which a

---

[2] See Polkinghorne (2005, pp. 30ff).

consistent and rich set of its possibilities of interaction has been manifested. For a physics in which time is just one coordinate variable that can in principle be considered as a totality, this is not too fantastic a notion. It might mean the return of final causality, in a new sense, to science. Only in the light of the manifestation of all the inherent possibilities of the universe, or at least of one set of compossible and extensive space–time states, might we be able to explain the properties of its originating simple parts.

We might think, as some quantum theorists do, of there being a set of possible states in phase space. The set of all possible such states would form an archetype of the possibilities for a universe. Instead of a wholly arbitrary set of ultimate laws and states that proceeds by wholly random processes to an unanticipated outcome, we might have a complete set of all possibilities, from which one set of consistent laws might be actualized. This set might include this space–time as one of many actualized states, or it might be the only consistently actualizable universe that contains intelligent agents like us. Mathematical physicists have proposed both possibilities.

Why, after all, should we think that the earliest and the simplest could provide a complete explanation of the later and more complex? Perhaps that idea belongs to an outdated mechanistic physics for which time is an absolute monolinear flow. Might we not think that the latest and most comprehensive state of a system, or the system taken as a whole, explains the simple origins? For the most comprehensive state would include the specification of all possible states, and a selection of actual states in terms of value ("value" being a notion that can be filled out in various ways). Then the laws of nature would not be wholly arbitrary principles of interaction. They would be principles necessary to the fruition of a coherent, complex, organized, and integrated universe of unique and inexponable value.

We could then speak of the supreme informational principle of the universe as the mathematically richest and most fertile set of states in logical space that could give rise to a physical cosmos that could be valued for its own sake. The set of all mathematically possible states (a set that would exist necessarily, and could not come into being or pass away) plus a selective principle of evaluation (a rule for ordering these states) would provide the informational code for constructing an actual universe.

That sense of information would be importantly different from the sense in which, for instance, DNA is a code for building bodies. It would precede, and not be the result of, any and all physical processes, evolutionary or otherwise. And it would not be part of the physical system for which it was a container and transmitter of information. But it would be analogous to "shaping" information, in that it would contain the patterns of all possible physical configurations, and a principle of selecting between possibilities.

If we cast around for some model for a non-physical carrier of information, containing patterns for possible existents, together with rules for ordering such patterns evaluatively, the historical example that springs to mind, or at least to the mind of any philosopher, is Plato's "World of Forms." This is precisely a world of archetypes in which the phenomena of the physical cosmos participate partially and imperfectly.

In some modern science such a Platonic model has proved attractive to mathematicians; and Roger Penrose, for one, has said that the Platonic realm is for him more real than the physical realm. It has a mathematical purity, immutability, and necessity that the observable physical world lacks. "To me," writes Penrose, "the world of perfect forms is primary . . . its existence being almost a logical necessity – and . . . the world of conscious perceptions and the world of physical reality are its shadows" (Penrose, 1994, p. 417).

Plato had difficulty in relating the world of forms (of possible states in phase space) to a dynamic power that could translate it into an actual physical embodiment. In Plato's dialogue *Timaeus*, the Demiurge or world architect uses the forms as models for constructing a universe, but seems strangely disconnected from the forms themselves (Plato, 1965, 29d–30d). It was Augustine, in the Christian tradition, who formed the elegant postulate that the forms were actually in the mind of God, necessary components of the divine being, which was the actual basis of their otherwise merely possible reality.[3]

13.2 Forms of consciousness

With the introduction of the idea of mind or consciousness as the carrier of possibilities, there is some motivation to move beyond the view that higher-level laws are just shorthand substitutes for boringly laborious lists of lower-level laws, and beyond the view that they are new principles of interaction between complex systems, the basic nature

---

[3] See Augustine (1991, Book 8, Chapter 2).

of the elements of such systems remaining what it always was. We may have to introduce the idea of consciousness as a distinctive kind of existent.

Consciousness is not just a new form of relationship between complex physical systems. Apprehension and understanding, and intelligent action for the sake of realizing some envisaged but not yet existent goal, are properties, not of physically measurable entities, but of a distinctive sort of reality that is not material.

If we posit consciousness as a distinctive kind of existent, we move to the third use of the term "information" – the semantic use, when some physical item (a written mark or sound) provides information about something other than itself to some consciousness that understands it. There are three main components here: the physical item, the person who takes it to refer or to indicate that some operation is to be carried out, and what it is about, or (in logic and mathematics, for example) the operation it instructs one to perform.

Digital computers operate in accordance with the second type of information. The computer is structured so that some of its physical components constitute a code for performing operations – there are physical elements with a function. But there is no one who understands the instructions; they operate automatically. Of course computer codes have been intentionally structured precisely so that the codes can be used for specific purposes, and the results on the screen can be understood by someone. That is the whole point of having computers. They are designed to help persons to understand things, and they provide information only when someone does understand what they produce.

Without that act of understanding, there is no information. There is only the material substratum that stores information – but that material basis needs to be interpreted by an act of intellectual understanding to become actual information.

That is why the "information" carried by DNA molecules is not information in the semantic sense. The code does provide a program for constructing an organism, but no person has constructed it and no consciousness needs to understand and apply the program. It has originated by ordinary evolutionary processes, and, like a computer program, it operates without the need for conscious interpretation.

Nevertheless, there may be a holistic explanation for the general process of evolution and for the sorts of organism that DNA codes construct. If we are looking for a total system within which "random" mutations and natural selection of specific kinds of organism occur, we might find in the ecosystem itself and its history a recipe for the generation of more complex physical systems and for the gradual development of organisms capable of conscious apprehension and creative response. Paul Davies and Simon Conway Morris are just two of the scientists who see in the basic physical foundations of the evolutionary process a vector to the virtually inevitable development of conscious and responsive life (Conway Morris, 2003; Davies, 1992).

It is extraordinary that a physical system generates informational codes for constructing complex integrated organisms. But that fact does not of itself require the introduction of any external designing intelligence. What is even more extraordinary is that these organisms then generate a quite new sort of information – semantic information – that does involve consciousness, interpretation, intention, and understanding.

In my view, such things as conscious intention and understanding have real existential status. They are irreducible and distinctive forms of reality. They are kinds of "stuff" that are not reducible to the properties of physical elements such as electrons. Yet they come into existence at the end of a many-billion-year-long process of development from simple physical elements.

If we are not simply to give up all attempt at explanation, and say that consciousness is just a random by-product of the evolutionary process, we must look for a different type of explanation: one to which contemporary biologists have largely been temperamentally averse, but which is now increasingly being forced upon our attention. That is, a cosmic holistic explanation, in which the development of the parts is explained by their contribution to the existence of an integrated totality.

Taken together, these considerations suggest the idea of a primordial consciousness that is ontologically prior to all physical realities, that contains the "coded" information for constructing any possible universe, and that can apprehend and appreciate any physical universe that exists. It would certainly be a strong reason for creating a universe that might contain finite consciousnesses that could share in appreciating, and even in creating, some of the distinctive values potential in the basic structure of the universe: for such a creation would increase the total amount and the kinds of value in existence.

Whether or not one calls such a primordial conscious-ness "God" is partly a matter of taste. For some, the idea of God is too anthropomorphic, too primitive and sentimen-tal, to be of use. But if some notion of value is introduced, as a reason for actualizing some rather than other logi-cally possible states, the notion of consciousness seems to be entailed. For it is consciousness that apprehends and appreciates value. Only intelligent consciousness can have a reason for bringing about some state, and that reason would precisely be the actualization and appreciation of some as yet merely possible value.

Consciousness, as a distinctive sort of real existent, not composed of purely physical elements, has been a major problem for classical materialism, and implausible attempts have even been made to deny that it exists at all. But quantum physics throws doubt on such denials. When quantum physics speaks of the collapse of a wave function when an observation is made, some quantum physicists hold that consciousness is involved in the actualization of possibilities in a constitutive way – as John Wheeler has put it, "It has not really happened, it is not a phenomenon, until it is an observed phenomenon" (1978, p. 14).

So for some physicists (and the list would be long, including John Wheeler, Henry Stapp, Eugene Wigner, John von Neumann, and Bernard d'Espagnat) con-sciousness is involved in the very existence of physical nature as it appears to us. Consciousness, as we know it, is capable of conceiving possibilities as well as apprehending actualities, and of making possibilities actual for a reason. Thus a hypothesis consonant with many interpretations of quantum physics is to see the actual world as rooted in a consciousness that conceives all possible states, and actualizes some of them for a reason connected with the evaluation of such states by that consciousness.

Such a reason might be that only one set of compossible states gives rise to a complex, interesting, and enduring universe – Leibniz's hypothesis (Leibniz, 1714, § 53–55) – or it may be that any universe can be actualized that exhibits a unique set of valued states, in which the values markedly outweigh the disvalues, and the disvalues are compensated in a way ultimately acceptable to those who have experienced them – Thomas Aquinas' hypothesis (Aquinas, 1265–1274, 1a, question 25, article 6).

The idea of holistic explanation is the idea of explaining the parts of an organic whole in terms of that whole itself and its fullest actualization. What is sometimes called "shaping information" is the property of some physical entities to store and transmit information, in the non-semantic sense of an ordered set of physical causes of more complex and integrated systems.

If there is a holistic explanation for the universe, it will explain its simplest laws and elements as preconditions of the realization of its fullest and most complex states. There is no doubt that the human brain is the most complex physical state so far known by us to exist. Consciousness and intelligent agency is generated by the central nervous system and the brain of *Homo sapiens* – and of course there may be further developments in knowledge and power yet to come, in other forms of organism, whether naturally or artificially produced. Rather as DNA may be seen as an informational code for constructing organisms, so the basic laws of physics – the laws of the interaction of complex as well as simple physical systems – can be seen as informational codes for developing societies of conscious intelligent agents out of simpler physical elements.

13.3 The supreme informational principle

However, the laws of physics did not, like DNA, evolve by mutation and selection, and they are not embodied in chemical or physical elements. Even those, like Lee Smolin, who speak of an "evolution" of physical laws, have to presuppose a prior set of laws that can account for such evolution. As a matter of logic, the laws in accordance with which physical entities relate cannot be generated by the relations between such entities. At least some basic set of laws must be seen as primordial and constitutive of reality rather than emergent from it.

My suggestion (it is actually the suggestion of many classical philosophers and theologians, and a suggestion that much modern physics supports rather than undermines) is that such basic laws can be fully understood only when they themselves are seen as preconditions for developing consciousness and intelligence from simple physical elements.

But then we have to see such conscious intelligence as a primary causal factor in the generation and nature of those simple physical elements. To adapt John Wheeler's suggestion a little, the simple originating phenomena of the universe may not even exist unless they are conceived, evaluated, and intentionally actualized by consciousness.

For some physicists, and I think for John Wheeler, it is the final conscious state of the universe itself that is a causal factor in its own physical origin. The universe generates a cosmic intelligence that then becomes cause of its own originating processes. But what this paradoxical suggestion really points to is the existence of a trans-temporal consciousness that can originate the universe as a condition of the existence of the sorts of consciousness the universe generates through and in time.

It has been objected that a consciousness cannot exist without some form of material embodiment, but this objection seems to rest simply upon a failure of human imagination. It is true that all consciousness requires an object; we are always conscious of something. But there may be many sorts of objects of consciousness. Human consciousnesses are fully and properly embodied, and their objects are normally physical, or at least sensory. But we can imagine, and even to some extent experience, consciousness of non-physical objects such as mathematical realities and unactualized logical possibilities. The cosmic consciousness being envisaged here would have the set of all possible universes as its object, and so it could not be part of any such universe (it may take embodied form in some universes, and Christians hold that it does, but it would also have to transcend any such form in order that those universes could exist in the first place).

In that respect, and unsurprisingly, cosmic consciousness is quite unlike any embodied consciousness. It is a primary ontological reality, in fact the one and only primary ontological reality, from which all universes are generated. This consciousness is the conceiver of all possible states and the actualizer of some, for the sake of values that are to be consciously apprehended and appreciated. This is the supreme informational principle for constructing universes.

Another objection that has been made, most publicly by Richard Dawkins in recent times (Dawkins, 2006), is that a cosmic consciousness is just too complex a thing to be likely to exist. The simple is more likely to exist than the complex, he says, and so to appeal to a cosmic consciousness is to try to explain the improbable in terms of the even more improbable, and that can hardly count as an explanation.

Something has gone wrong here with the use of the idea of probability. It is false that the simple is more likely to exist than the complex – there are infinitely more complex possible states than simple states, and so, if anything, a complex state is more likely to exist than a simple one. But of course no single possible state is either more or less likely to exist than any other possible state. Probability does not really work when considering the likelihood of anything at all existing. Considerations of probability alone cannot tell us what is likely to exist, out of the complete array of all possible states of affairs.

13.4 Combining nomological and axiological explanations

Some think that the fact that anything exists is ultimately just a brute fact for which there can be no explanation. But there are two general sorts of explanation that are widely accepted and that may together suggest an explanation as to why a universe exists.

One is the nomological explanation generally used in the natural sciences, by which appeal to a general law and an initial state makes the existence of some further physical state necessary. The other is the axiological explanation used in the human sciences and in human life generally, by which appeal to motives or reasons ("I did X in order to get Y") makes the existence of some state intelligible.

Nomological explanation gets stuck when it comes to explaining why the laws of nature are ultimately as they are. Many physicists would like to see something necessary about the fundamental laws of nature, so that they could not be otherwise and could not fail to exist. But what could that be? A possible suggestion is that they could be necessary in the sense that they are conditions of (necessary to) realizing a set of distinctive values (reasonable goals of action). Those values in turn would be necessarily what they are if there is a complete array of possible states that can generally be ranked in order of value. So if we can think of an array of all possible states that could exist, with the values they necessarily have, there would be an intrinsic reason for the existence of any universe: namely, the goodness that it would exhibit.

A combination of nomological and axiological explanation, of necessity and value, suggests the idea of a complete set of possible states, a set that would be necessary in that there is no possible alternative to it. All such states would have degrees of goodness necessarily attached to them. Some of these would, by necessity, be negative – that is, they would be disvalues or evils. All possibly actualizable coherent universes might be such that it would not be possible to eliminate all evils from them. But some would have higher degrees of value than others, or perhaps different kinds of incommensurable values worth having. So there would be an internal reason for the selection of some such states for existence.

372

We are operating at a level of great abstraction here, but my main point can be made simply. If there is no ultimate reason for anything existing, then it is not true that the simple is more likely to exist than the complex. But if there is an ultimate reason, it would have to lie in the goodness or value of certain possible states that are necessarily what they are.

This is, and is meant to be, a basically Platonic idea – one that has been revived in recent years by, among others, John Leslie and Roger Penrose (Leslie, 1989; Penrose, 1994). I have suggested, following Augustine, that mind or consciousness is somehow involved in such an ultimate explanation, because it is mind that stores possibilities non-physically, and mind that can act for a reason. This is just to say that mind is a fundamental constituent of ultimate reality, and is necessarily prior to all physical entities. For they are actualizations of possibilities apprehended by cosmic mind, the only actuality that is not capable of being brought into being or of not existing or of being other than it is, as it is a condition of the existence of all possibilities whatsoever. Cosmic consciousness is the condition of any and all possibilities existing (which they necessarily do), and not merely a very complex thing that just happens to exist.

It is clear that any such "Platonic" view cannot accept that information is necessarily materially embodied, as the primary informational source, God, is not material. But it may still be the case that human consciousness is materially embodied, and that it is not simply something quite different in kind from material objects, as it lies in an emergent continuum with material entities that have no consciousness.

13.5 Human embodied consciousness

Human consciousness is oriented towards what can be known by the senses, and it is embodied in a material world that provides both the sources of its information and the arena for its intelligent responses. Human information also needs to be stored in a material form, in language and in physical areas of the brain. It is not something purely mental or unembodied.

The material elements in themselves provide no information, however; they are the carriers of information, and without them no information is carried. But the material stuff needs to be interpreted by someone to denote some thing or process. Consciousness needs material objects with which to operate. It uses such objects in two ways – to form an organized informational storage system, and to prompt acts of intellectual understanding.

That is why what is often, and perhaps not quite fairly, called Cartesian dualism is an inadequate account of consciousness. For it gives the impression that there are parallel worlds of pure unverbalized ideas in the mind, and of words and physical objects that somehow "image" or copy such ideas in physical form – the mind is not only the mirror of nature, but human language is a mirror of totally non-physical relations of ideas that already fully exist in the mind.

The "parallel worlds" idea of mind and body misses the necessity of physical objects and sensory data for the mind, and it also misses the necessity of a conscious mind if those physical objects are to be, in a proper sense, information: objects that refer beyond themselves, that "mean something," when interpreted by a socially trained and historically situated understanding.

Human languages are the vehicles of semantic information. They do not correspond to some realm of pure ideas that humans just tune into. They are culturally distinct and develop historically and differently by use and practice. Human minds learn some such language, and that largely governs how they think and what they think about. But "understanding" is a distinctive capacity that can learn and develop a language, that can use language creatively, and appreciate the products of such creativity.

There are here three distinctive capacities of the human person, unique among all organisms on Earth, so far as we can tell – the capacity to be sensitive to and appreciative of information received, to be creative in responding to it, and to learn and develop such capacities in relation to other persons in specific historical contexts. Human persons receive information, interpret it, and transmit it in a fully semantic way.

Humans nevertheless stand in a continuum that begins from the much simpler capacity of physical objects to respond to stimuli from an environment of other objects. The registration of the stimulus, the largely automatic response, and the form of interaction with other objects, are elementary forms of what becomes, in humans, conscious apprehension, creative response, and personal relationships with other persons.

Because this continuum exists, we can use the term "information" to apply at various stages. Even the simplest physical object "registers information" from its environment, "interprets" it, and acts on the basis of it – but of course none of these simple capacities involves consciousness or awareness. There is nothing there that is truly creative, and there is no development, as there is with human persons, of a unique historical trajectory, no sense of an inward spiritual journey or a novel and unpredictable history.

As organisms become more complex and integrated, these primitive capacities of registration and response are extended and become more diverse and individual. Consciousness seems to be a continuously emergent property that is so closely integrated with organic systems that it may seem right to call it an emergent aspect of a monistic and naturalistic system, as Arthur Peacocke did.[4] It is as well to remember that even the notorious Descartes said, "I am not just lodged in my body like a pilot in his ship, but I am intimately united with it, and so confused and intermingled with it that I and my body compose, as it were, a single whole" (Descartes, 1637, p. 161).

Antipathy to Descartes is today so strong that some writers even miss out the "not" in this quotation, thus changing its sense completely. Mind and body are, for Descartes, a "single whole," and thus the alleged founder of dualism seems paradoxically to espouse a form of monism. However, this is a double-aspect monism (the philosopher Charles Taliaferro (1994) more helpfully calls it "integrative dualism"), and the two aspects can, however improperly in the case of humans, be torn apart. So it is possible to have a consciousness without a body – God is such – and it is possible to have a functioning brain

---

[4] See Chapter 12 of this volume.

without consciousness (although we assume this does not normally, or perhaps ever, happen).

Material embodiment is more than contingent for humans. As Thomas Aquinas said: if souls exist without bodies after death, they do so "in an unnatural and imperfect way" (Aquinas, 1265–1274, I, question 76, article 1). Humans are fully embodied minds. Yet in human consciousness an important threshold is crossed to full semantic information, and that suggests the idea of ultimate reality as a consciousness that holds the information necessary to create any universe, the ultimate ontological and informational principle.

It must be kept in mind that if this is to be more than an interesting hypothesis, it must have some experiential impact. Religion, ambiguous though it is, aims at its best to promulgate disciplines of mind that can relate humans to the cosmic consciousness that it sees as compassionate and perfectly good. It is important to bear in mind that religion does not depend on the success of some speculative theory about the universe. It depends upon the sense of human beings that they can apprehend a personal reality that is other and better than they, and that supports and encourages their own strivings for goodness.

But such a sense of apprehension of transcendent goodness needs to be supported by a general view of reality that is coherent and plausible, and within which an idea of transcendent goodness has a central place. Precisely because our views of reality must be informed by scientific knowledge, theologians must engage with science in formulating metaphysical theories that, however tentative, show religious commitment to be reasonable and intellectually appealing.

Classical materialism may be dead, but naturalistic views of the universe are very much alive, and one of the great challenges for naturalistic thinkers is to provide an adequate account of the remarkable role played by information in our current understanding of the physical world. I do not think any responsible theorist would say that this has been done. However, great strides have been made in recent years, and there is little reason to say we know it is impossible in principle.

My suggestion, however, has been that most uses of the term "information" rely for their significance upon being analogous to, and logically depend upon, the primary sense of "semantic information." This, in my opinion, is not accidental, for it is in that sense that one may hold a view of the universe as constructed on an informational pattern that is carried and transmitted by the mind of God. The God hypothesis is not contradicted by any, and is quite strongly supported by some, of the speculations of contemporary information theory. So my conclusion is that the ultimate ontological reality is indeed information, but that information is ultimately held in the mind of God, and such a hypothesis expresses one of the most coherent and plausible accounts of the nature of ultimate reality that is available to us in the modern scientific age.

13.6 Conclusions

Two types of information have been discussed: "shaping" information and semantic information. For the former, information is a code for the construction of complex integrated systems, and is best understood by holistic (whole–part) explanation. This may be seen either as a "shorthand" explanation for complex cases, or as involving new laws governing the behavior of emergent complex systems. Cosmic holistic explanation seeks to explain parts of the cosmos in terms of its total structure and history. This suggests the idea of an ultimate informational principle for the universe – a set of all possible states in phase space, and a rule for ordering them in terms of value. Such a principle would be logically prior to and ontologically different from any actual physical state.

A Platonic–Augustinian model for such a principle is the "World of Forms," an ultimate informational system carried and transmitted by a cosmic mind. This is a fully semantic sense of information, for which data are understood and interpreted as significant by consciousness. Such a cosmic consciousness is, or is part of, what has been called God in classical Christian theology.

For some quantum physicists, consciousness is essentially involved in the actuality of any observable phenomenon. On the theistic hypothesis, there is one cosmic consciousness that is essentially involved in the actuality of any universe. It carries complete information about all possible states in phase space, states that carry necessary evaluative rankings, and thus provide an internal reason for the existence of one or more actual universes. This would provide an ultimate explanation for the existence of our universe.

Human consciousness lies on an emergent continuum with primitive and non-conscious stimulus–response entities, and it is by nature embodied. The mind–body relation in humans can best be termed double-aspect monism or integrative dualism. Human consciousness must have sensory content and a physical means of functioning and expression. But it is not the only form of consciousness. The notion of semantic information is extensible to cover the idea of a cosmic, unembodied consciousness, which carries and transmits the informational code for the construction of this and any possible universe. That is the mind of God.

References

Aquinas, T. (1265–1274). *Summa Theologiae*, Latin text and English translations, vols 1–60, ed. Thomas Gilby (1964–1966). London and New York: Eyre & Spottiswoode and McGraw-Hill.

Augustine (c. 400–416). *De Trinitate*, trans. E. Hill (1991). New York: New City Press.

Conway Morris, S. (2003). *Life's Solution*. Cambridge: Cambridge University Press.

Davies, P. (1992). *The Mind of God*. New York: Simon and Schuster.

Dawkins, R. (2006). *The God Delusion*. London: Bantam Press.

Descartes, R. (1637). *Discourse on Method*, trans. A. Wollaston (1960). Harmondsworth: Penguin.

Leibniz, G. W. (1714). Monadology. In *Discourse on Metaphysics and the Monadology*, trans. G. R. Montgomery (1992). Buffalo, NY: Prometheus Books.

Leslie, J. (1989). *Universes*. London: Routledge.

Penrose, R. (1994). *Shadows of the Mind*. Oxford: Oxford University Press.

Plato (1965). *Timaeus*, trans. H. D. P. Lee. Harmondsworth: Penguin.

Polkinghorne, J. (2005). *Exploring Reality*. London: SPCK.

Puddefoot, J. C. (1991). Information and creation. In *The Science and Theology of Information*, eds C. Wassermann, R. Kirby, and B. Rordorf. Geneva: Labor et Fides.

Taliaferro, C. (1994). *Consciousness and the Mind of God*. Cambridge: Cambridge University Press.

Wheeler, J. A. (1978). The past and the delayed-choice double-slit experiment. In *Mathematical Foundations of Quantum Theory*, ed. A. R. Marlow. New York: Academic Press.

14

Information, theology, and the universe

JOHN F. HAUGHT

~

The most important single issue in the conversation of theology with science is whether and how God acts in or influences the world. Here I shall ask whether the notion of information can help theologians address this question. It is well known that traditional philosophies and theologies intuited a universal "informational" principle running through all things. Their sense that "Mind," "Wisdom," or "Logos" inhabits and globally patterns the universe has been repeated in widely different ways time and again: in ancient Greek philosophy, the Wisdom literature of the Hebrew Scriptures, Philo, early Christianity, Stoicism, Hegel, Whitehead, and others. But can the intuition that the universe is the bearer of an overarching meaning – of an informational principle actively present to the entire cosmic process – have any plausibility whatsoever in the age of science?

These days, after all, one must hesitate before connecting the Logos of theology immediately to patterns in nature. The life process as seen through the eyes of evolutionary biologists, to cite the main reason for such reluctance, scarcely seems to be the embodiment of any

Information and the Nature of Reality: From Physics to Metaphysics, eds. Paul Davies and Niels Henrik Gregersen. Published by Cambridge University Press

universal divine principle of meaning or wisdom. Contrary to the picture of cosmic order expressed in much religious thought, evolution involves seemingly endless experimentation with different "forms," most of which are eventually discarded and replaced by those only accidentally suited to the demands of natural selection. The impersonal Darwinian proliferation of experimental life forms, only a few of which seem to be adaptive for any length of time, scarcely reflects anything like an underlying divine wisdom. The spontaneous origin of life, the apparent randomness of genetic variation that helps account for the diversity of life, and the accidents in natural history that render the trajectory of the whole life story unpredictable, make one wonder just how "informed" the natural world can be after all. Certainly evolution makes the hypothesis of divine *design* questionable. If anything, nature seems, at least on the surface, to be the product of what Richard Dawkins (1986) calls a "blind watchmaker." How, then, can theology think coherently of a divine presence meaningfully operative amid the blind impersonality and aimless contingency manifested in science's new pictures of nature?

14.1 Information as analogy?

At the very outset, such a project must acknowledge that theological discourse about divine action is inseparable from the language of analogy. Theological ideas cannot and should not be translated into scientific propositions, as attempted by "creation science" and intelligent design theory. Whenever theological speculation emulates scientific precision in specifying how divine action occurs – for example, by locating it in the hidden realm of quantum events – there is the risk of rendering mundane that which radically transcends nature. God, as theologian Paul Tillich insists, cannot be merely one cause among others in the world and still be appropriately spoken of as God (Tillich, 1967, p. 238). So analogical language is indispensable in all attempts to understand divine action, providence, wisdom, and purpose. However, the fact that theology must use analogy is not something for which the theologian ever needs to apologize, as though clarity and quantitative precision would be more appropriate. Analogy (as well as symbolic expression in general) is essential to protecting the subject matter of faith and theology from being reduced to what is objectifiable in the form of "clear and distinct ideas." Because faith is a matter of being grasped by, rather than of grasping, that which one takes to be "Ultimate Reality," its primal language can never be cured of the vagueness associated with symbol, metaphor, and analogy. The further theology drifts from the figurative discourse of its original inspiration, and the more closely its expression imitates the explanatory models of science, the more it loses touch with its own depth.[1]

Nevertheless, if in theology analogies are essential, some are less suggestive than others. For example, the image of God as a "designer" has become increasingly questionable, especially in view of evolutionary accounts of life. Is it possible, then, that the notion of "information" may be less misleading than that of design in theology's inevitably tentative reflections on how divine purposive action could be operative in the natural world?

---

[1] On faith as a matter of "being grasped," see Tillich (1958, pp. 76–85).

14.2 Information and improbability

In the broadest sense, "information" can mean whatever gives form, order, pattern, or identity to something, whether it be an electron, a crystal, the human mind, a civilization, or an economic system. It is similar to what Aristotle meant by formal cause. In a general sense "form" is a principle of limitation that gives specificity to things. "Things" could not even be actual unless they have a definite pattern or form. Comparably the idea of "information" as used in communications theory means the reduction or removal of indefiniteness or uncertainty.[2] The more uncertainty is removed, the more information there is. It is in this sense that I shall explore the analogical suggestiveness of the idea of information for theological discourse. I shall be asking whether the universe that sponsors evolution, even though it fails to fit comfortably with the analogy of design, can nonetheless be viewed intelligibly as "informational," and hence as consistent with a divine purposive presence. Information, as John Bowker insists, does not just slop around blindly in the universe. It has to be channeled and processed by an information *system* (Bowker, 1988, pp. 9–18, 112–143). I want to ask then whether the universe itself could be thought of theologically as something like an information system through which a "message" of ultimate importance is being communicated.[3]

Every information system is limited, capable of handling only a finite amount of information. Hence care

---

[2] See Campbell (1982, p. 255).
[3] Here the "universe" includes humans and their own creations.

must be taken to encode a message in such a way as to fit within the boundaries of what is allowed by the medium. As there is the additional possibility that a message will be obscured or lost in static or other kinds of interference, it must be encoded so as to be heard over and above the scrambling effects of "noise." So "redundancy" is a necessary aspect of information processing (Seife, 2006, pp. 5–20). Redundancy lowers the economy and speed of a communication, but it compensates for the delay by ensuring accurate transmission of meaning. On the one hand, by overdoing the redundancy of a message, one risks reducing it to the merely probable. Too much redundancy can cause information to lose its informative edge, so to speak. It would inhibit the emergence of novelty. On the other hand, if one avoids redundancy altogether a message may get lost in the noise.

In order for a message to come across as informative it must include novelty. It must appear as contrasting with a hypothetical background of merely probable messages. According to the informational analogy that I am following here, the *amount* of information in a message varies in direct proportion to the improbability of its content. If I anticipate that all the messages from a given source will be the same, then when one of them is delivered it carries little or no information as I already know what it is going to say. There is an apparent paradox involved here: the more informative a message is, the less immediately comprehensible it may be, at least as far as the communication of meaning is concerned. If so, then a maximally informative message, if we could conceive of the finite universe as its carrier, would be accompanied by the least redundancy. It would be something like that which cosmologists call a singularity, and as such inaccessible to conventional science, as scientific method in its concern for predictability is most at home in the realm of what appears probable or predictable rather than improbable or unpredictable. So a message might be unreadable by conventional science to the degree that it is improbable – that is to say informative.[4]

This means that science would be inadequate to saying anything one way or the other about the supposed reality of that which is singularly informative. Human consciousness would have to adopt some other, perhaps non-scientific, mode of receptivity in order to attune itself to what is ultimately informative. Such a "message" would lie beyond any ordinary or scientific capacity to grasp. It may be so novel, and "improbable," that it would escape the grasp of objectifying consciousness altogether. We could be grasped by it, but we could not grasp it.[5]

If the universe is the bearer of a religiously revelatory meaning, therefore, such a meaning would not fit neatly into any standard scientific way of understanding. It may be so improbable as to be completely ignored by science. Science (at least conventionally) is incapable of dealing

---

[4] Perhaps, however, the idea of information, along with the related notion of complexity, is changing subtly and gradually the nature of scientific understanding. On this point see both Campbell (1982) and Seife (2006).

[5] Accordingly, theology would associate the idea of God or revelation with novelty and not just order. This is true of St Paul referring to the Christian life as a "new creation," of Paul Tillich's reference to the principle of redemption (presented in the picture of Jesus as the Christ) as "New Being," and of process theology's identifying God as the principle of novelty that inevitably disturbs the status quo. God is the "one who makes all things new."

with the uniqueness of improbabilities, but instead seeks to reduce the improbable to the probable, and hence requires the redundancy of many instances of similar occurrences in order to formulate general laws and theories. Any information about divine action or cosmic purpose in the universe would not show up on screens that are wired only to receive what is predictable and probable. And so it would make sense that the most significant information carried by the cosmos transcends scientific understanding altogether.

14.3 Noise, redundancy, and revelation

Theological inquiry into how divine action may be involved in the unfolding of the universe, including the evolution of life, may carry the informational analogy further by looking more carefully at the ideas of noise and redundancy that are closely associated with the idea of information in communication theory. "Noise," for example, suggests contingency, chance, accident, or randomness. At the level of living phenomena, noise is correlative to the unsystematic waywardness and wildness characteristic of the contingencies in life and natural history that apparently prod the life story onto indeterminate trajectories. Ever since Darwin, the sense of contingency in evolution has challenged the theological claim that there could conceivably be a divine cosmic "design" for life in the universe.[6] However, the high degree of accident in evolution can become theologically intelligible once it is

---

[6] One of the best examples of this view of contingency can be found in the work of the late Harvard paleontologist Stephen Jay Gould. See, for example, Gould (1977, 1989).

interpreted in terms of an informational universe. If the universe is in some analogous sense an information system, then it would allow the unfolding of occurrences within it to make their way between the extremes of noise on one side and redundancy on the other. Contingency in the cosmos, and especially in life's evolution, therefore, would not rule out the hypothesis that the course of cosmic events, at a level too deep to be detected by science, may be influenced by a creative, providential informing agency. The true character of such a course of events would be inaccessible to science, inasmuch as it could not be mapped without remainder onto the efficient or material kinds of causation that science is interested in clarifying mathematically. Yet, it could still be deeply informative, and hence revelatory, nonetheless.

Furthermore, if the cosmic process carries in its deepest dimensions a theologically relevant content, the informational analogy allows us to assume that the "redundant" presence of deterministic, predictable habits (such as the laws of physics or the "mechanism" of natural selection) do not render the universe completely opaque to the hypothetically informing influence of God. Modern scientific naturalism has sometimes assumed that nature is a closed continuum of physically deterministic causes and effects, and hence completely impermeable to any possible divine action or revelation. It is for this reason that Albert Einstein rejected the idea that authentic religion can properly include belief in a personal, responsive deity (Einstein, 1954, p. 11). However, the habituality of nature, if we follow the informational analogy, does not require such an extremist claim. For what appears in nature to be deterministic routine is simply a mental (mathematical) abstraction of only one aspect of a richer informational process that concretely speaking winds its way through time between the extremes of noise and redundancy.

The real world, in other words, is a blend of order and indefiniteness. The informed, but unfinished, universe of which I am speaking here is a process involving the ever-new ordering and reordering of relative disorder. There is a sense in which all information is the ordering of indefiniteness. In the sentences I am reciting right now, for example, the relative indefiniteness of a figuratively disassembled set of characters of an alphabet is the raw material that is being assembled by the writer into a specified informational pattern. An entirely arbitrary series of letters is being given a non-arbitrary form in my sentences and paragraphs.

In order to function as potential bits of information, the letters of the alphabet must have a random, "noisy" nature. That is, the characters must be capable of being figuratively disassembled and placed in an imaginary "mixing pot," whence they can be called out, one by one, and placed in a non-arbitrary informational sequence. Inherent in any communication through the medium of a code is a randomizing feature, or an entropic tendency, that functions as the necessary condition of being reassembled into novel informational patterns. Put otherwise, a code has to have the capacity to disassemble in order to reassemble. Suppose, for example, that the English alphabet were locked deterministically and irreversibly into a single sequence, so that b always has to follow a, c always has to follow b, d always has to follow c, and so on. This would be an ordered entity but one with very low informational content. As long as it remains impossible to break down such rigidity, the communication of written information would be impossible. In other words, too much order – or design – would prevent the transmission of information. If the universe or life were simply "designed," it would be frozen in a fixed and eternally unchanging identity. Design is a dead end. Its rigidity would prevent the entrance of emergent novelty. Absolute order would be antithetical to any genuine cosmic emergence, as everything would be fixed in frozen formality.

This informational truism is especially interesting theologically today because the dismissal of "God" by modern evolutionary naturalists, and the rejection of evolution by creationists and "intelligent design" devotees, is often a result of their shared observation that the Darwinian world fails to conform to simplistic notions of design and order. There is no need here to enter into a thorough discussion of creationism and intelligent design. It is well known that their antipathy to evolution is rooted in the assumption that a world filled with accidents or contingencies is too noisy to be rendered compatible with their idea of a designing deity. But it is also worth noting that the atheistic ideas of the renowned evolutionist Richard Dawkins and many other evolutionists are also based on the same assumption: namely, that any God deserving of the name would also have to be a designer in the same sense as "intelligent design" proponents understand the ultimate cause of living complexity. In his obsession with the dead-horse of design, Dawkins insists that any reasonable affirmation of God's existence would require that living organisms exhibit perfect engineering. So Dawkins' implicit theological assumptions are

391

essentially identical to those of his "intelligent design" opponents.[7]

Reasonable theology, however, has long ago done away with the simplistic identifying of God with a "designer," or divine action with design. Even before Darwin had published the *Origins*, Cardinal John Henry Newman, to give a noteworthy example, insisted that theology has little use for Paley's brand of natural theology with its focus on design. Design-oriented theology, he insisted, could "not tell us one word about Christianity proper." It "cannot be Christian, in any true sense, at all." Paley's brand of "physical" theology, Newman even goes on to say, "tends, if it occupies the mind, to dispose it against Christianity."[8]

In place of design, I would suggest that natural theology may more appropriately understand divine influence along the lines of informational flow, although this too is an analogy that can never adequately capture what actual religious experience understands as ultimate reality. The point of considering the informational analogy at all is that, unlike design, it is fully open to the fact of nature's contingency. It does not insist that the plausibility of the idea of God depends upon the existence of ordered perfection in the natural world. Intelligent design proponents along with Dawkins, on the other hand, consider the fact that nature and life are speckled with accidents or spontaneities to be contrary to divine action and cosmic

[7] In his most recent bestseller *The God Delusion*, Dawkins (2006) defines God as "a superhuman, supernatural intelligence who deliberately designed and created the universe and everything in it, including us" (p. 31). My objection here is not to Dawkins's atheism but to his theology.

[8] Here citing J. H. Newman (1959, p. 411). Newman is also famous for saying that he believes in design because of God, not in God because of design. See the discussion of Newman by McGrath (2005, pp. 29f).

purpose, and hence as support for atheism. However, the existence of a God "who makes all things new" is, at least informationally speaking, consistent with the existence of an abundance of accidents in evolution.

Moreover, the informational character of nature meshes naturally with the idea that nature is narrative to the core. One of the consequences of developments in geology, biology, and cosmology during the last two centuries is that the universe now manifests itself as an unfolding story rather than an essentially fixed state of being. Underlying this narrative character, however, lies the more fundamental fact of the universe's informational make-up. Only by way of a relatively "noisy" disassembly would the cosmic process allow, at least occasionally, for the emergence of newer narrative patterns. Thus the universe's entropic veering toward disorder may be essential for its unfolding as a meaningful story. Some degree of "noise" is a necessary aspect of divine creativity or revelation rather than its perpetual enemy. If a hypothetical divine informational principle is in any analogous sense expressing itself in the unfolding of cosmic events, it would not be surprising that natural process constantly harbors a reservoir of indefiniteness in order to have both a real future at all and the opportunity to give birth to elaborate instances of order such as that displayed in the emergent phenomena of life, mind, and culture. The cosmic oscillation between noise and redundancy is parallel to the encoding of information from letters of an alphabet. Without nature's capacity for moments of deconstruction no evolutionary "story" could be inscribed in it. Without a constant inclination toward a state of noise the universe could not be the carrier of a meaning.

Hence the fact of random, undirected occurrences in life's evolution, or for that matter in the wider cosmic process, is not decisive evidence that the universe lies beyond the pale of providence or divine purpose, wisdom, and compassion. Nor is the classic Einsteinian sense of the absolute inviolability of nature's "laws" proof of the imperviousness of the cosmos to purpose. Ever since the nineteenth century, a whole generation of scientifically educated cosmic pessimists have become accustomed to fixing their attention on either entropy or predictable physical routines as though either of these constituted direct evidence of the pointlessness of the universe.[9] Naturalists have observed correctly that any ordered system tends to lapse into disorder. But they have often failed to understand that any truly interesting *story* requires contingencies that let in enough novelty to overcome sheer redundancy. They have failed to entertain the possibility that even if the universe as a whole is headed toward death by entropy, in the meantime something momentous may nonetheless be working itself out narratively here and now.

If the element of contingency in natural processes can be rendered theologically intelligible in terms of the notion of information, so also can that of redundancy (or what is misleadingly called "necessity"). The incarnation

[9] "All that science reveals to us," says Cornell historian of science William Provine, "is chance and necessity... Modern science directly implies that the world is organized strictly in accordance with mechanistic principles. There are no purposive principles whatsoever in nature. There are no gods and no designing forces that are rationally detectable. The frequently made assertion that modern biology and the assumptions of the Judeo–Christian tradition are fully compatible is false" (Provine, 1989, p. 261).

of purpose in the cosmos would require an element of redundancy such as that embedded in the orderly and "monotonous" routines of physical law. Too much novelty of expression would interfere with the communication of information. So it is theologically significant that information requires predictable order along with the improbability of novelty. Accordingly, an informational universe would have to wend its way narratively between the two extremes of absolute noise and absolute redundancy.

Overlooked in much modern and contemporary cosmic pessimism – a world view that has assumed the overall pointlessness of the universe – is the simple fact that the communication of meaning inevitably requires the breakdown of present order, the dismantling of temporary instances of rigid design, if the universe is to sustain its informational and narrative character. Cosmic pessimism feeds parasitically on the uncalled-for apotheosis of either noise or redundancy, both of which can easily be abstracted and contrivedly isolated from each other and from the concrete temporal process that weaves the two elements together informationally. One wing of cosmic pessimism absolutizes the trend toward chaos as fundamental. Exaggerating nature's stochastic aspects, it then "explains" the precarious instances of emergent order such as a cell or a brain as momentary reversals of the general trend of the universe toward energetic collapse.[10] Meanwhile, the other wing of cosmic pessimism is so entranced by the recurrence of physical law that it views emergent phenomena as nothing more than "simplicity masquerading as complexity" (Atkins, 1994, p. 200).

[10] For example, Monod (1972).

I believe that both types of pessimism, though they claim to be the very epitome of realism, are illogical and unscientific. In Whitehead's terms they are both instances of the logical "fallacy of misplaced concreteness" (Whitehead, 1925, pp. 51, 58). Scientifically speaking, moreover, both wings of cosmic pessimism are anachronistically fixated on the early modern obsession with mechanical design as paradigmatic for the understanding of nature, and hence also as the only appropriate model for understanding divine action. Consequently, after failing to discover any indisputable evidence of flawless divine engineering in nature, they declare the cosmos to be devoid of meaning or purpose.[11]

14.4 Information and story

In the cosmic process increments in the intensity of information are essential factors in the evolution of more complex forms of life and consciousness over the course of time. Increase in information is the fundamental ingredient in the phenomenon of emergence, and it is the motive force in evolution. As the cosmos has passed through its evolution from matter to life, and then to consciousness, ethics, and civilization, something new, and hence relatively improbable, has been added at each emergent phase. This novelty is real, and not just an illusory cover-up of what is taken to be at bottom either absurd "chance" or blind "necessity."

But what exactly constitutes the novelty at each later and higher level so that it cannot be explained exhaustively

---

[11] Monod's thought flits back and forth between one type of pessimism and the other. See his book *Chance and Necessity* (Monod, 1972).

in terms of the earlier and simpler states of physical existence? Classically minded scientists are puzzled as to exactly what the emergent reality is. As the total amount of matter and energy remains constant throughout all transformations, it may appear to a mechanistic mindset that there is never *really* anything new, but instead merely a reshuffling of atoms and molecules, in the evolution of the later and higher levels. There may even be an objection to my use of terms such as "higher" and "lower,"[12] as these adjectives are inapplicable to what seems to be a purely impersonal set of purely material processes.

We can see now, however, that the novelty in an emergent universe is analogous to what is now called information. Information can continue to settle into our world without in any way altering the laws of thermodynamics. And just as information can be given to my word processor without in any way modifying the rules governing the lower levels in the computational hierarchy, the insinuation of novel forms of information into the fabric of the universe happens so unobtrusively that it often goes unnoticed by limited scientific sensibilities.

The introduction of new information into the universe takes place in such a way that each later and higher emergent development relies upon earlier and lower levels without in any way violating the "laws" operative at the earlier and lower levels. The passing on of genetic information, for example, does not suspend or alter the laws of chemistry and physics. So it is a logical mistake to suppose that one can discern or measure the increments of such

---

[12] Here the terms "higher" and "lower" refer to distinct levels of complexity.

emergent information as though it were on the same logical and ontological level as that studied by the sciences of chemistry and physics. The level of information is hierarchically distinct from the level being informed.

However, informational novelty requires as a condition of its emergence a base of subsidiary probability and redundancy. This is clearly the case in both biological processes and verbal communication. In the latter case the often unnecessary repetition of certain words and sounds can make our message come across more clearly to the average listener. Analogously in the case of life, if it were not for the repetitive and redundant routines of chemical activity and recurrent organic assemblies, the "improbable" living cell could not emerge or be sustained in existence.

Moving up to a higher level, the even more improbable occurrence of human thought could not take place without relying on the relatively redundant subsidiary neurological and physiological processes that serve as its substrate. Emergent novelty in the cosmos is impossible apart from the reliable functioning of "monotonous" subservient routines. Such redundancy is especially evident in the mammalian brain, where in order to guarantee the integrity of perceptive and mental processes nature has endowed animal and human brains with an incredibly extravagant quantity of "circuits." For example, compared with a computer – in which a single wire is sufficient to activate a "gate" – thousands of nerve fibers may empty into a single neuron in the brain.

However, redundancy unchecked can also be a hindrance to communication. Its function is to prevent the "noise" of disorder from drowning out the flow of information. But it can become so pervasive at times that it inhibits the birth of new patternings. So redundancy must be broken through if genuine novelty is to emerge. If redundancy reaches the point of absolute inflexibility, it degenerates into a monotony that is antithetical to the emergence of information. Yet without at least some redundancy, new degrees of cosmic complexity could never emerge. The surprise in emergent novelty requires physical redundancy as the scaffolding by which it sticks out from what is merely probable.

The informational blending of order and novelty with deep time may well be the stuff of a truly momentous, although still unfinished, *story*: one that may carry a meaning far outstretching our ability to comprehend things locally and scientifically. At the very least, in any case, information is a much more flexible notion than "design" as a vehicle of meaning. Thus it would be more appropriate theologically to speak of a *narratively informed*, rather than a mechanically designed, universe. Information allows theology to think analogously of divine action as occurring somewhere between the extremes of absolute randomness on the one side and complete redundancy on the other. But it does not require that there be no deviations from design. If we view the cosmos as informational, we will not be surprised that there is disorder at the margins of all organization (as the first creation account in Genesis already intuits). The notion of "design," by contrast, is intolerant of any such disorder. And where disorder is forbidden, novelty is also excluded, and along with it any notion of a truly living God.

The real world, as well as any world that could narratively embody a meaning, will inevitably oscillate between the monotony of excessive order on the one hand and seemingly meaningless chaos on the other. The risk entailed in any truly informational universe is that it may deviate at times from the "appropriate" mix. My point, however, is that the alternatives – either a world of sheer contingency or a world deadened by mechanical routine – would be incapable of carrying any meaning whatsoever. Hence the idea of "intelligent design" not only fails to fit the requirements of science, but it also fails theologically by its embarrassment about the contingency in nature. It fails to see that the universe can be deeply informational without being designed. However, at the other extreme, modern evolutionist materialism – with its declaration that any universe that sponsors Darwinian processes is inherently meaningless – is no less unwarranted. This belief system, one that is becoming more and more respectable in the academic world and intellectual culture today, is in fact the result of a logical fallacy that confuses contingency with absolute noise on the one hand, and law with absolute redundancy on the other. Here the *actual* cosmic blend of novelty and order gets lost in mutually segregated abstractions that are in turn illogically mistaken for concrete reality.

Concretely, the universe is a *blend* of order and novelty, steering its way between absolute redundancy on the one hand and absolute noise on the other. It is intuitively evident from ordinary experience that information must walk the razor's edge between redundancy and noise. It should not surprise us, therefore, that any universe that carries a meaning would permit its content to meander adventurously between the two extremes of deadly design and unintelligible chaos. Hence any meaning that our universe may be carrying would go unnoticed both by a science that looks only at what is predictable, and by a tragic "realism" that expects the final cosmic state to be the complete triumph of chaos. A different kind of reading competency, one cultivated by the virtue of hope, may be necessary if we are to discern any ultimate meaning expressing itself in the cosmic process. Even if we were to be grasped by such a meaning, however, we could not realistically expect full understanding until the story is complete. At present the universe's deepest intelligibility can be dimly discerned, if at all, only if we turn our minds – and hearts – toward the cosmic future. Adopting such a posture of hope, of course, is a thoroughly religious challenge, but it is not one that contradicts science, especially if we think of the cosmos in informational terms.

14.5 Information and evil

Finally, however, our universe, even if it is meaningfully informed, is still one in which great evil can exist alongside good. Can the informational analogy be of any help to theology as it deals with this perennial problem? I believe so, if we can agree with Alfred North Whitehead and process theologians that there are two distinct kinds of evil in the universe, and if we view it as a still-unfinished creative process rather than just a static collection of entities.[13] Information straddles the two domains of chaos and order, both of which at their extremes would inhibit the flow of information. The real world has to be a mixture of harmony and nuance. It is the adventurous nature of such a

---

[13] See Cobb and Griffin (1976).

world that it is subject to deviating at times from what we humans may consider the "appropriate" blend, either by being too monotonously ordered or too carelessly disordered. Our ordinary experience tells us that information must walk the narrow ridge between too much order and too much chaos. If the universe is in any way something like an information system, it too would allow its content to manifest itself between the two extremes. Any information processed by the universe could easily be eclipsed by excessive chaos or deadened by too much order.

Excessive redundancy and unnecessary noise constitute analogies for the two distinct types of "evil" that occur in the realms of life and human experience. The evil of excessive redundancy consists of the endless repetition of routines when the introduction of novel information would make way for the emergence of more being and value in the cosmic process. At the human level of cosmic emergence, an example of the "evil of redundancy" might be the obsession with our own cognitional certitude and existential security to the point that we ignore the political, cultural, scientific, or religious complexity of the world. It would take the form of a resistance to novelty and adventure, which are required to prevent the decay of human life and civilization. On the other hand, evil can also take the form of unnecessary noise as well. The "evil of noise," considered in the context of a cosmology of information, consists in an excessive disregard for rules of order without which the carrying of meaning becomes impossible.

The presence of these two types of evil in the world does not logically contradict belief in the reality of divine creation and revelation. For if we emphasize the self-giving and non-coercive nature of an informing God, we should not be surprised that cosmic evolution and human existence would often take an "erring" and meandering course, slinking alternatively toward either noise or monotony, but over the long haul perhaps sketching a truly informative meaning between the extreme instances of each. Information does not require that there be no deviations from order at all, as entailed by the idea of "divine design." Consequently, if we could learn to view the cosmos as an information system, and cosmic becoming as an informational process, we should not be utterly surprised that at least some degree of disorder and redundancy would show up at the margins of its development. Unlike the idea of information, the notion of "design" is intolerant of disorder, and wherever disorder is completely ruled out so also is novelty. And wherever novelty is excluded so too is a truly interesting universe, as well as a proportionately challenging sense of ultimate reality and meaning.

References

Atkins, P. W. (1994). *The 2nd Law: Energy, Chaos, and Form.* New York: Scientific American Books.

Bowker, J. (1988). *Is Anybody Out There?* Westminster, MD: Christian Classics.

Campbell, J. (1982). *Grammatical Man: Information, Entropy, Language and Life.* New York: Touchstone.

Cobb, J. B., and Griffin, D. R. (1976). *Process Theology: An Introductory Exposition.* Philadelphia: Westminster.

Dawkins, R. (1986). *The Blind Watchmaker.* New York: W. W. Norton.

Dawkins, R. (2006). *The God Delusion.* New York: Houghton Mifflin.

Einstein, A. (1954). *Ideas and Opinions*. New York: Bonanza Books.

Gould, S. J. (1977). *Ever Since Darwin*. New York: W. W. Norton.

Gould, S. J. (1989). *Wonderful Life: The Burgess Shale and the Nature of History*. New York: W. W. Norton.

McGrath, A. (2005). A blast from the past? The Boyle Lectures and natural theology. *Science and Christian Belief*, 17: 1, 25–34.

Monod, J. (1972). *Chance and Necessity: An Essay on the Natural Philosophy of Modern Biology*, trans. A. Wainhouse. New York: Vintage Books.

Newman, J. H. (1959). *The Idea of a University*. Garden City: Image Books. Philadelphia: The Westminster Press.

Provine, W. (1989). Evolution and the foundation of ethics. In *Science, Technology, and Social Progress*, ed. S. L. Goldman. Bethlehem, PA: Lehigh University Press.

Seife, C. (2006). *Decoding the Universe: How the New Science of Information Is Explaining Everything in the Cosmos, from Our Brains to Black Holes*. New York: Viking.

Tillich, P. (1958). *Dynamics of Faith*. New York: Harper Torchbooks.

Tillich, P. (1967). *Systematic Theology*, vol. I. Chicago: University of Chicago Press.

Whitehead, A. N. (1925). *Science and the Modern World*. New York: The Free Press.

15

God, matter, and information: towards a Stoicizing Logos Christology

NIELS HENRIK GREGERSEN

~

Up to modernity, the majority of Christian thinkers presupposed the world of creation to be composed of two parts: the material and the spiritual, existing alongside one another as independent yet interacting realms. In the traditional exegesis of Genesis 1, for example, the creation of light in Genesis 1:3 ("Let there be light") was interpreted as a spiritual light for spiritual beings in a spiritual world (*kosmos noētos*), preceding the creation of the corporeal light of the sun in the empirical world (*kosmos aisthētikos*) in Genesis 1:14.

This two-stock universe lost its plausibility with the advent of classical physics in the seventeenth century, when nature came to be seen as a seamless unity. The scientific intuition of the oneness of the universe, however, was initially combined with a narrow interpretation of the nature of the material. As Isaac Newton (1642–1727) argued in his *Opticks*, matter is basically atomic: "solid, massy, hard, impenetrable, moveable

Information and the Nature of Reality: From Physics to Metaphysics, eds. Paul Davies and Niels Henrik Gregersen. Published by Cambridge University Press © P. Davies and N. Gregersen 2010, 2014.

particles".[1] According to Newton, these particles, formed in the beginning by God and held together by the mechanical laws of nature, serve the divine purpose of the universe while at the same time being embraced by God, who is ubiquitously at work ordering, shaping, and reshaping the universe. For Newton, mechanism and theism were two sides of the same coin. How else to explain the orderliness of the otherwise arraying particles? God, the creator of the world of matter and the author of the deterministic laws of nature, was continuously providing the collaboration and ends of all biological creatures:

the Instinct of Brutes and Insects, can be the Effect of nothing else than the Wisdom and Skill of a powerful ever-living Agent, who being in all Places, is more able to move the Bodies within his boundless uniform Sensorium, and thereby to form and reform the Parts of the Universe, than we are by our Will to move the Parts of our own Bodies.   (Newton, 1952, p. 403)

How else to explain the growth and appropriateness of biological organs? And finally, despite the physical determinism, there was supposed to be room for the causal efficacy of moral and spiritual powers of human agents. Human duty towards God and human beings "will appear to us by the Light of Nature," as Newton

---

[1] "All these things being consider'd, it seems probable to me, that God in the Beginning form'd Matter in solid, massy, hard, impenetrable, moveable Particles, of such Sizes and Figures, and with such other Properties, and in such Proportion to Space, as most conduced to the End for which he form'd them; and that these primitive Particles, being Solids, are incomparably harder than any porous Bodies compounded of them; even so very hard, as never to wear or break in pieces; no ordinary Power being able to divide what God himself made one in the first Creation." (Newton, 1952, p. 400)

concludes his *Opticks* (1730, p. 405). Hence, there are not
only laws of nature, but also a natural law guiding the
human deliberation about right and wrong.

As time went on, however, the philosophical doctrine of
classical materialism, reigning in the eighteenth and nineteenth centuries, made the further claim that the world of
material particles is the sole reality, and that all genuine
information about nature and humanity therefore must
be reduced to the causal powers inherent in the interplay between basic physical constituents. Julien Offray
de La Mettrie's *L'homme machine* (de La Mettrie, 1748)
exemplified the programmatic reduction of the soul to
the mechanics of the human body, just as Pierre Simon
Laplace's *Exposition du Système du Monde* (Laplace, 1813)
a little later epitomized the conviction that the universe
at large is a closed physical system of interacting particles,
leaving no job to be done for a god (apart, perhaps, for
initiating the world system). According to these versions
of materialism, matter replaces God as ultimate reality,
while the mental world is excluded from the inventory
of genuinely existing things. For what is real evidences
itself through causal powers, the innate powers of material objects.

15.1 Neutral monism and the irreducible
aspects of matter

The options for theologians of that day were either to
oppose materialism radically, as the traditionalists and
philosophical idealists did, or to dike materialism by
claiming that the choices of human freedom, guided
by ethical principles and sentiments, cannot be causally

reduced to natural movements, as proposed by Immanuel Kant (1724–1804). Assuming a split between nature and culture, the Kantians – friendly, but disinterestedly – handed over nature to scientific determinism. Leaving ontological questions aside as "speculative metaphysics," the so-called liberal theologians subsequently occupied themselves with axiological questions of ethics and aesthetics. Human experiences of beauty, love, and moral sensibility, they said, are themselves pointers to a transcendent reality that is more ultimate than the deterministic order of natural events.

Reflective theologians of today seem to have more intellectual options. For as physicists Paul Davies and John Gribbin have put it, during the twentieth century the matter myth was broken, for "Matter as such has been demoted from its central role, to be replaced by concepts such as organization, complexity and information" (Davies and Gribbin, 1992, p. 15). Another way of phrasing the new situation is that the concept of matter has been significantly enlarged so as to include the stuff-character of matter (as evidenced in quarks, electrons, atoms, molecules, etc.), the energy of matter (the kinetic potential and changeability of physical matter), and the informational structures of matter (its capacity for pattern formation). The distinction between the aspects of mass and energy took place with Albert Einstein's special theory of relativity (1905), which showed that mass and the energy of motion are equivalent in a quantitative sense without being simply identical. Matter comprises both energy and "rest mass" (see Ernan McMullin in Chapter 2 of this volume). In the wake of this discovery, philosophers of science began to discuss whether relativity theory eventually demands a "dematerialization" of our inherited concept of matter. For example, in today's cosmology entirely new forms of "matter" are hypothesized; "dark matter," for example, has a mass responsible for gravitational attraction but without emitting any radiation.

The next blow to the matter myth came with the quantum theory of Niels Bohr and colleagues in the 1920s, in which material particles lost quite a few of their former "primary qualities," such as simple localization, duration, and indivisibility. Finally, since the end of the 1940s, the new sciences of cybernetics (Claude Shannon, Norbert Wiener, John von Neumann, and others) and biological information theory have shown that informational properties of matter seem to exert a specifiable causal influence, and hence should be seen as irreducible aspects of the material world. In short, the new picture is that matter is not just the kind of physical brick-like stuff that Newtonian phycisists used to think of, and that *mass, energy, and information constitute three irreducible though inseparable aspects of the material world.*[2] A possible rejoinder here is that information differs from the mass and energy aspects of matter by not being quantifiable in a manner comparable to the mass units of *grams* and the energy units of *joule*. And indeed, as we shall see below, the term "information" takes different meanings depending upon whether we define each individual quantum event as an informational event (a qubit, as suggested by Seth Lloyd in

---

[2] *"Der Begriff der Materie hat drei Hauptaspekte, die untereinander untrennbar, aber relativ selbständig sind: der stoffliche (mit Beziehung zum Substrat), der energetische (mit Beziehung zur Bewegung) und der informatorische (mit Beziehung zur Struktur und Organisiertheit)"* (Zemann, 1990, p. 695).

Chapter 5 of this volume), as a computational digit (Shannon information), as patterned information (Aristotelian information), or as meaning information (semantic information). However, as information in all these forms seems to play a causal role in the history of nature, there are strong reasons for giving information a central role in a scientifically informed ontology. First, information could be said to be at the bedrock of physical reality in terms of the generative capacities of quantum processes. "Information" is here simply what generates differences. And second, informational structures play an undeniable causal role in material constellations, as, for example, in the physical phenomenon of resonance, or in biological systems such as DNA sequences. "Information" is here about differences that make a difference in the story of evolution. So, just as *informational events* are quintessential at the bottom level of quantum reality, so *informational structures* are at work as the driving forces for the historical unfolding of physical reality (see Chapter 1 of this volume).

Within contemporary theology, coherent ways of understanding God as ultimate source of all-that-is, and the world of creation as based in physical processes, have been presented.[3] One such version (that I am following) works on the basis of a "neutral monism," which accepts the principle that all-that-is-and-will-evolve in the world of creation is based on a "natural" substrate with some structuring capacity, while not identifying this substrate with particular sets of physical descriptions thereof. The point here is not prematurely to eliminate the causal capacities of higher-level properties, such as information

---

[3] See, for example, Peacocke (2007).

and intentionality, as part of the comprehensive picture of what is ultimately real. The term "neutral monism" was used by both William James and Bertrand Russell to indicate the view that the world is one in origin and enfoldment ("monism"), while admitting that we cannot do justice to the complex evolution of natural systems by only using the physical terms of mass and energy (hence "neutral" with respect to particular physical theories of matter).[4] The agnosticism implied in the term "neutral" monism acknowledges our inability to describe ultimate reality from one particular perspective (say, "mass," "energy," or "information"); neutral monism thus favors a pluralistic set of explanations of reality, although under the serious constraint that any viable monism must include what we have come to know as basic physical terms. No informational structures, no mental events, and no human agents emerge unsupported by mass and energy, and no flow of information happens without a suitable physical basis. And yet, the one world as we know it through the sciences as well as from everyday experience is multifarious. The sort of neutral monism that I wish to defend is thus a "multiformed monism."

15.2 Balancing Platonizing and Stoicizing tendencies in Christian theology

How can contemporary religious reflection deal with the multifaceted concept of the material that came out of twentieth-century physics? Elsewhere (Gregersen, 2007)

---

[4] See Stubenberg, 2005; and my own previous use in Gregersen (1998).

I have argued that theology should take a strong interest in scientific suggestions of a comprehensive concept of matter as a field of mass, energy, and information. The Christian idea of a Triune God – Father, Son, and Holy Spirit – may even be seen as a preadaptation to later developments, and hence be a unique resource for developing a relational ontology that is congenial to the concept of matter as a *field* of mass, energy, and information. Already from the fourth century onwards, the Cappadocian Fathers (Gregory of Nyssa, Gregory of Nazianzus, and Basil the Great) developed a concept of divine nature and life according to which God is not conceived of as a single entity or person (as suggested by later seventeenth-century theism), but as a *community* of persons or interacting poles. The being of God is here not taken as a pre-established substance. Rather the incomprehensible divine nature (*ousía*) and the consistency of divine generosity and love result out of the co-determinative actions between three divine centers of activity: Father, Son, and Spirit. The "Father" is the ultimate source of divine life and the existence of cosmos, the "Son" or the Logos is the formative principle in God, and functions also as the informational resource of creation, while the "Holy Spirit" is the divine energy that also energizes the world of the living.

There is not only a formal parallel between the three irreducible modes of divine being and the triad of matter as mass, information, and energy. From a theological perspective, there must also be an ontological connection between God and world by virtue of the incarnational structure of the doctrine of creation: God is present in the midst of the world of nature as the informational principle

(Logos) and as the energizing principle (Spirit). Only the originating principle of the Father remains consistently transcendent, thereby responding to the metaphysical question: Whence the universe? Accordingly, the idea of a divine Logos answers the question: Whence the informational resources exhibited in the history of the universe? Finally, the idea of the Spirit answers the question: Whence the energy and unrest of natural processes? It is only in the interplay between information (Logos) and energy (Spirit) that the world of creation produces evolutionary novelties rather than mere repetitions (Gregersen, 2007, pp. 307–314). On this view, of course, what is ultimate reality from a physical point of view is penultimate from a theological perspective. A theological ontology will thus assume that God is intimately present at the core of physical matter as described by the sciences (and beyond that), without thereby conflating God the Creator and the world of creation.

The conception of God as Creator, hence as the ultimate source of all material processes, points to a remaining Platonizing element in all kinds of theism (here broadly conceived of as including the Trinitarian view). The notion of God's transcendence retains the religious understanding of God's ontological priority (as Creator) and of the consistency of God's self-determination to love and generosity (as Redeemer and Fulfiller) during the ups and downs of temporal flux. A co-terminous logic, however, demands that God as Creator should be conceived of as actively creating within, through, and under the guise of material processes (where else in this monist world?). This points back to an often-forgotten Stoicizing tendency within Christian belief, especially with regard to the central Christian belief in God's incarnation in time–space.

In what follows, I am going to discuss how the concept of information as arising from the perspective of physics and biology may inform the idea of a Logos Christology, which is not confined to the historical figure of Jesus, but has universal scope from the outset. But I am also going to work the other way around: How can a Christian theology *redescribe*, from the specific perspective of a Logos Christology, the idea of information as already described and (partially) explained by the sciences? The point of a theological redescription here is not to establish an inferential argument from the irreducible aspect of information in the material world to the "existence" of a divine principle of Logos. I am not entertaining a "natural theology" based on science. Rather, I am propounding a theological hypothesis that claims that the theological assumptions of a Logos Christology are highly congenial with basic assumptions of the understanding of matter and the material that came out of twentieth-century scientific developments. As such, my theological reasoning is working under both scientific and philosophical constraints. First, the theological candidate of truth that the divine Logos is the informational resource of the universe would be *scientifically* falsified if the concept of information could be fully reduced to properties of mass and energy transactions. Second, its *philosophical plausibility* would be significantly reduced if the actual outcomes of the interplay between energy and information during evolution would be so pointless that the religious assumption of a generous and caring creator would be existentially counter-indicated. "All work, and no play, makes the universe dull," to rephrase an old saying.[5]

15.3 Mass-and-energy aspects of matter

Already in the mid nineteenth century the concept of matter became more comprehensive and less solid than hitherto assumed. Eventually, the concept of energy was taken to be equally important, or even more fundamental, than the concept of mass. In his "Remarks on the Forces of Inorganic Nature" (Mayer, 1842), German natural philosopher Julius Robert Mayer formulated a principle that pointed forwards to a fundamental change in the scientific concept of matter. The essential property of force or energy, according to Mayer, consists of "the unity of its indestructibility and convertibility" (Mayer, 1980, p. 70). A little later, in 1851, English physicist William Thomson (the later Lord Kelvin) intimated: "I believe the tendency in the material world is for motion to become diffused, and that as a whole the reversion of concentration is gradually going on" (Thomson, 1980, p. 85).

Thereby the intuitions were formed that were later formulated in the first and second laws of thermodynamics. The first law of thermodynamics states that energy is conserved when its energy is put into work and converted into heat; heat appeared to be a general property of matter. In 1865 Rudolph Clausius then formulated the second law of

---

5 "All work, and no play, makes Jack a dull boy." I don't know the exact provenance of this proverb, but it is famously featured in the movie *The Shining*, in which Jack Nicholson repeatedly types the phrase.

thermodynamics, stating that energy exchanges are irreversible. In a closed system, a portion of energy converted into work dissipates and loses its force to do the same work twice. Thus, the energy is at once a constant feature of matter and a more and more inefficient capacity of matter, as time goes on. Understanding the universe as a closed system by assuming the first law of thermodynamics, the law of entropy predicts the bleak perspective that the universe is going to be less and less capable of producing the heat necessary for living organisms to survive.

With Albert Einstein's general theory of relativity (1916), the concept of energy was further generalized by understanding mass and energy as quantitatively equivalent. In a vacuum, energy (E) is numerically equal to the product of its mass (m) and the speed of light (c) squared: $E = mc^2$. At a closer look, this famous formula can have two different philosophical interpretations (both compatible with a neutral monism). It can mean that "mass" and "energy" are two equal properties of an underlying material system, or it can be taken to mean that energy and mass constitute the same stuff, which then appears with different emphasis in different systems. In some systems, the mass-aspect of matter dominates, while at other places, matter takes the form of a field. In Einstein's and Infeld's *The Evolution of Physics*, the latter view is expressed, as follows: "Matter is where the concentration of energy is great, field where the concentration of energy is small" (Einstein and Infeld, 1938, p. 242).[6] This distinction between matter and field especially highlights the fact that most matter is not visible in any form, but can be

[6] See Flores (2005, pp. 4–6).

evidenced only indirectly by its gravitational force. Eventually the old concept of matter being composed of solid material particles became obsolete with the time–space and energy–matter unities of relativity theory. The new situation was clearly perceived by Bertrand Russell:

Matter, for common sense, is something which persists in time and moves in space. But for modern relativity physics this view is no longer tenable. A piece of matter has become, not a persistent thing with varying states, but a system of inter-related events. The old solidity is gone, and with it the characteristics that, to the materialist, made matter seem more real than fleeting thoughts.   (Russell, 1961, p. 241)

Even more revolutionary for the concept of matter was quantum mechanics, which gave up the idea of material entities having a simple definable state at the ultimate level of matter. The uncertainty principle of quantum mechanics suggests that particles emerge out of and perish into a field of subatomic events with an ontological status that defies description in terms of simple location or duration. The stochastic nature of the decoherence of quantum events into classical events and the continuous entanglement of otherwise distant events give evidence that there is no scientific basis left for the common-sense view of matter. Matter has become a deep but elusive concept. Atoms are not undividable entities, as the old etymology of *a-tomos* (Greek for "un-divided") suggests. Nor are atoms, according to entanglement, always separated from one another according to the Humean scheme of disparate causes and effects.

Discussing the impact of relativity theory and quantum mechanics, philosopher Norwood Russell Hanson has even spoken of the "de-materialization" of the scientific concept of matter: "Matter has been dematerialized, not just as a concept of the philosophically real, but now as an idea of modern physics... The things for which Newton typified matter – e.g., an exactly determinable state, a point shape, absolute solidity – these are now the properties electrons do not, because theoretically they cannot, have" (Hanson 1962, p. 34). Hanson's point, of course, is not that physical events do not have a material basis, but that the concept of matter is undergoing serious revisions. The visibility, indivisibility, and locality of old-style materialism have gone.

Also biology poses new questions for materialism by bringing information into the center of attention. Despite all claims of causal reduction, physics (not only classical, but also modern) has failed to explain basic features of biological evolution. For whereas the properties of some chemical compounds (for example, those known from the chemical table) are formed under specific circumstances as a result of atomic affinities fully explainable by physical laws, there exists no law for the sequences of DNA macromolecules. Thus, genomes are arbitrary relative to underlying chemical affinities. Genomes are as they are because of contingent historical circumstances. But if DNA sequences are causally efficient instructors by virtue of their informational structure, information can no longer be left out of a comprehensive picture of what drives nature. For again, what is causally effective, is real. As a matter of fact, information co-determines how organisms make use of their available energy budgets.

15.4 Information is about differences

The question here arises: What is meant by information? In what follows I assume that information, taken in a generic sense, has to do with the generation and proliferation of differences. In this sense, information lies at the root of material existence in the form of quantum events, which incessantly produce differences when potential quantum fields decohere and become realized as individual events. This is the main thesis of Seth Lloyd (2006), discussed in Chapter 5 of this volume. However, most quantum events cancel each other out so as to produce a rather uniform result. Only where quantum events not only produce differences, but differences that make a causal impact with long-lasting effect, do we have "a difference that makes a difference" (to use Gregory Bateson's famous term). Finally, when we approach phenomena of salient difference, we have "differences that make a difference for somebody," who evaluates a difference as important or salient in some or other respect.

By defining information by the production and proliferation of differences, we can go back and differentiate John Puddefoot's helpful distinction between three types of information (Puddefoot, 1996, pp. 301–320). The first is *counting information*, or the mathematical concept of information as defined by Claude Shannon. Here "information" means the minimal information content, expressed in bits (*b*inary dig*its*), of any state or event: 1 or 0. In order, for example, to know which of 16 people have won a car in a lottery, one would expect to need 16 bits of information, namely the 15 losers (symbolized by a 0) and the one winner (symbolized by a 1). However, if one began to ask intelligently (is the winner in this or that group of 8, and so on), one would only need 4 computational steps ($\log_2(16) = 4$).[7] Here "information" means mathematically compressible information. This quantitative concept lies at the heart of computational complexity studies. The question is, however, whether this mathematical concept of information has an ontological status. At first sight, it does not. Shannon information is not an ontological theory about the world per se, but a procedure about how to analyze chunks of information using the minimally necessary steps, and how to transmit that information. However, there must also be some ontological basis for the mathematical theory of information. Not only is any information state always embodied in a concrete physical medium, but this physical medium also determines what Terrence Deacon (in Chapter 8 of this volume) calls the "information potential," that is, the very capacity of physical media for carrying and storing information. No information without an information apparatus (a calculator, a digital machine, a human brain). Secondly, ontological issues come to the fore as soon as Shannon information is used to achieve information *about* something ("information" in the semantic sense), say: "Find the winner of the car." Here the information system refers to physical background conditions distinct from, and absent from, the informational system itself. In the example above, the background assumption is the equiprobability of winning the car among the 16 potential winners (that is, a situation of high entropy in Boltzmann's physical sense), plus the

---

[7] See Weaver (1949, p. 100f).

one "winner" who has not yet been singled out. Imagine, however, that we had a background context of inequality; say, the information that the winner happens to wear a moustache. In this case, we might find the winner in fewer computational steps, starting with step one: "Identify men with moustache." We would then be able to find the winner in one step (if there were only one with moustache), in two steps (if there were two), in three steps (if there were up to three or four). As formulated by Deacon, whereas the Boltzmann entropy concerns the possibility that an informational signal will be corrupted, the Shannon entropy of a signal is the probability of a given signal being present in a given physical context. In short, the informational concept of entropy cannot be reduced to the physical one, as the former, when used *about* something, specifies something as standing in the foreground of attention ("the potential winner"), while countless other aspects of the physical situation are deemed irrelevant (say, the armchairs in the room, or the interaction between molecules).

Different from Shannon information is what Puddefoot calls "shaping information," which is the form or pattern of existing things. Here the interest lies in morphology, the study of forms, or specific characteristics. Shaping information may derive either from internal sources (such as a zygote) or from external constraints that cause something to have a definite pattern in relation to its environment. This shaping information is what Shannon information is *about* when used to acquire knowledge about the environment. But notice that shaping information is pervasive in the world surrounding us, depending on all kinds of boundaries that could fall under observational interest, as well as on the scale of investigation. For example, is the snail and its house one shape, or a compound of two? I therefore suggest making a further distinction within the general category of "shaping information." Shaping information seems to have two forms, either in form of the mere production of differences (as in quantum events), or in the form of larger-scale, semi-stable or resilient structures in the classical domains of physics, chemistry, and biology. In the next section, I refer to these two forms of shaping information as "cutting information" and "channeling information," respectively.

The third type of information is what we refer to in daily parlance: coming to know something of importance. In *meaning information*, information is not only about something, but is of interest *for somebody* in a given context. An approaching military helicopter can be a neutral fact of no importance, a salient sign of a fatal combat, or a long-awaited rescue. In meaning information, we are not only interested in the *fact* that there is this or that feature in our environment (to which we may refer), but we are interested in what it *means to us*. Information here is part of a process of communication, as pointed out by Bernd-Olaf Küppers (in Chapter 9 of this volume).[8]

---

[8] Admittedly, it is still an issue of debate whether semantic meaning is always part of a communication process in relation to agents, or whether semantic information makes up a logical realm of its own, as suggested, for example, by Karl Popper. I thank Dr Torben Bräuner at Roskilde University for raising this question (email of 13 May 2009). But I leave the discussion here, since my interest is on the potential causal effect of meaning information, a causal role that presupposes some recipients for whom a given information is of importance. For biological agents, new meaning information usually elicits new actions with world-transforming effects.

As argued by both Deacon and Jesper Hoffmeyer (in Chapters 8 and 10 of this volume), the aspects of salience and aboutness may already be present at biological level. However, wherever we identify the entrance of meaning information (at basic cellular level, at brain level, or at the level of human communication) we cannot slide easily from one aspect of information to another. Novel informational features emerge during the course of evolution. It seems to me, however, that *the concept of shaping information is basic in relation to the two other forms of information*. Counting information seeks to model or compress shaping information (as derived from either quantum or classical sources). Meaning information enters into the picture whenever biological agents take interest in their own future. The capacities of such agents, however, are themselves based in higher-order structures of the shaping kind. Meaning information comes about when interpreting shaping information for a specific purpose.

The question is now whether it is possible to identify different types of shaping information from physics to biology. This is a vast task beyond my competence. Let me therefore identify just a few relevant types of shaping information that are of importance for philosophical and religious reflection.

15.5 From quantum information to biological information: cutting and building up

In *Programming the Universe* (2006), Seth Lloyd sets out to present a picture of the universe as a universal computer. The universe, however, is not *like* a digital computer, as suggested by Stephen Wolfram (2002). The universe *is* a computer, but a quantum computer whose bits are qubits (quantum bits). "Every molecule, atoms, and elementary particles register bits of information. Every interaction between those pieces of the universe processes that information by altering those bits. That is, the universe computes" (Lloyd, 2006, p. 3). The universe began computing from its start, and what it computes is itself.

My interest here does not concern Lloyd's challenging thesis of a self-computing universe, which involves the view that the universe literally (not only metaphorically) "registers" itself.[9] For my purpose, the generative capacity of quantum events to *produce differences* is important. According to Lloyd, "Information and energy play complementary roles in the universe: energy makes physical systems do things. Information tells them what to do." This view also involves that "the primary actor in the physical history of the universe is information" (ibid., p. 40).

In a minimalistic reformulation of Lloyd's thesis, this means that each and any quantum event not only *does* something on the basis of the immediate situation of the universe (performing an energy transaction), but also, by its occurrence, *instructs* (informationally) the immediately following situation in which other quantum events are going to occur. Since the relation between two quantum events (A and B) is not unilaterally preset from A to B

---

[9] The term "register" might seem to suggest both an ability to store information, and to observe one's behavior, which is hardly the case for quantum processes. In a personal communication with Seth Lloyd (Copenhagen, 17 July 2006), he clarified that the term "registering" should be understood like "making a step," without any memorizing or observational allusions.

(although the overall relations between As and Bs take place within the statistical limits of quantum mechanics), qubits do not behave, and do not make instructions, in the same manner as a digital computer, which runs its programs on a classical hardware system. Qubits do not behave like binary digits, where, in each computational step, there are only two possibilities (o or 1), one of which is predetermined by the software program (when x, do o, when y, do 1). The behavior of qubits often has many more possible outcomes.

While the processing of information on a digital computer is in principle foreseeable (although often totally unexpected from a practical point of view), the processing of qubits is in principle unforeseeable. One could therefore never be able to make an exact *replica* of the quantum history of the universe on a technological quantum computer. Seth Lloyd's proposal, however, involves the idea that much less could do the work for us. Assuming that the universe at large *is* a computer (in the sense of doing minimal steps and instructing the conditions for future steps), the universe is *indistinguishable* from a quantum computer (ibid., p. 54). That is, if we technologically can establish local quantum computers, using qubits and not digits and working almost as fast and as efficiently as quantum processes themselves, we could analyze chunks of quantum processes bit for bit. As a consequence, our knowledge of the real-world informational processes would be manifold increased, and we could begin to make inferences from local qubit behaviors-and-instructions to a more general picture of the cosmic processes at large. I am intrigued by this proposal, and I believe this view of reality has interesting implications.

(1) Quantum events, "the jiggling of atoms," *make decisions* for the future universe via quantum decoherence.

(2) The universe is at the bottom level of qubits a *generator of differences*; hence of informational novelty.

(3) Accordingly, *information tends to grow* with the history of the universe.

(4) *Entropy is reinterpreted* as a measure for the "invisible information" (not as a loss of information); that is, the information that we are not able to harvest (although the information necessarily exists because quantum events continue to occur and generate differences).

(5) Since the universe is indistinguishable from a quantum computer by making steps and instructing further steps, the universe is *continuously computing* itself, whatever the consequences may be for large-scale systems (which is of concern to living beings).

Thus, the information we are dealing with here is indifferent to meaning information. Accordingly there is no specific storage of the "interesting" forms of information. The universe as a quantum computer just goes on computing in a stochastic manner, without concern for outcomes. In this sense, one might say that the language of the quantum machinery is like saying "cut-cut": that is, carving out specific physical pathways under specific pressures while eliminating possibilities.

This view changes if we enter into the classical domains of thermodynamics. Although quantum differences mostly even each other out when entering the medium-sized world, some do indeed have causal effects (as quantum chemistry repeatedly shows). Causal amplifications must have their first-time instantiation in quantum processes, but once the higher-order systems have come into being many examples of amplification also take place in classical domains, owing to the fact that classical systems tend to build on previously established structures. The formation of crystals offers one such example of amplification processes; tornados another. Self-organizational structures emerge by keeping up the delicate balance between positive amplification and negative feedback. In the context of specific boundary conditions, pathways are carved out with far-reaching historical consequences. History and topology begin to matter.[10] *Hence, the language of pattern-formation via amplification and self-organization is like "build on, build on." We are no longer only dealing with stochastic "cutting" information, but with historically oriented "channeling" information.* Historically stable forms of information are beginning to take over.

This development reaches a new level in the world of biology, when a strong chemical storage of information takes place in the macromolecules of DNA, and when the distinction between what is inside to an organism and what is outside to an organism begins to matter. In his programmatic article (in Chapter 7 of this volume), John Maynard Smith points out that information is a quintessential concept in biology, because DNA – like culture – is concerned with the storage and transmission of information. He argues that there is a fundamental difference between the *genes*, which are "codes" that instruct the formation of specific proteins, and the *proteins*, which are coded for by genes. Even if proteins themselves often have a strong causal effect as inducers or repressors of genes, their causal

---

[10] See Deacon (2003, pp. 280–284).

role is arbitrary in relation to the genes. Thus, there is no necessity about which inducers regulate which genes; it is a question of happenstance or *gratuité* (as Monod called it). By contrast, the genes code for something very specific to take place, such as the production of eyes. Genes are not only informational in the sense that they *store information* about a biological past; genes are also informational in the sense of *providing instructions* for adaptive purposes. In this sense, argues Maynard Smith, genes are not only of central importance for molecular biology, but also for developmental biology. Genes are intentional, not in a mental sense but in the sense of seeking desirable outcomes. This element of post-hoc "intentionality" is itself a result of natural selection.

This view clearly gives DNA a special status, because DNA instructs for specific outcomes rather than just working in a haphazard way. This biochemical concept of information is unilateral in orientation. Genetic information "codes" for the epigenetic processes of proteins, while there is no structural feedback from the developed cell to the genetic instruction, other than the subsequent feedback on differential survival. By contrast, the biosemiotic approach, represented by Jesper Hoffmeyer and others, wishes to extend the informational perspective to the level of the cells (Emmeche, 1999). Any cell "interprets" its environment by using its available resources in accordance with the "interests" of the cell. As put by Hoffmeyer, the "inventory-control system" of the DNA always "goes through a user interface" (Hoffmeyer, 2008, p. 166). This interface is provided by the cell in its immediate context (usually the organism itself); the "habits" of a cell's conduct are thus the product of a long and intertwined evolutionary history. Thereby the specific role of the genes as carriers of information and as instructors for protein formation need not be challenged. The biosemiotic proposal is not Lamarckian, if we by Lamarckianism understand a notion of DNA-based heredity of acquired characteristics. But biosemiotics assumes a bigger part to be played by the "interpreters" of the genetic information: that is, *the local organisms*, acting under environmental possibilities and pressures. The behaviors of cells and organisms are themselves a product of a long ("Baldwinian") learning history (Deacon, 2003).[11]

It remains to be seen whether the biosemiotic approach will be able to point to forms of causality that cannot be explained by more standard biomechanical approaches (such as chemical bondings and attachments). However, from a philosophical point of view, the biosemiotic approach has the advantage of explaining how the emergence of meaning information, or biological "interest," can come about, so to speak, on the shoulders of the instructional information of the genome. Although the biosemiotic approach presupposes the notion of instructional information in the well-known biochemistry of genes and proteins, it places the gene story in the wider context of the life of the basic units of the organism: the cells.

---

[11] The so-called "Baldwin effect" is named after the Darwinian psychologist James Mark Baldwin (1861–1934), who famously argued for the idea of "organic evolution." According to Baldwin, intelligence (and its biological proto-forms) play a positive adaptive role, as when individual organisms – by way of habits or learned imitation – accommodate to their environments by selective emphasis and response; see Richards (1987, pp. 401–503).

Expressed in continuation of the metaphors used above: the formation of the biological "habits of interest" is not only about cutting information (which generates differences for evolution to work on), nor just about instructional information (which builds up protein structures), but also about an information of connectivity (of absorbing and bringing into resonance a given situation with the interest of biological systems), from cells to organisms. As we are now going to see, all three aspects of information are central to Christian theology.

15.6 Information and Logos Christology: theological perspectives

In what follows, my aim is to explore how the scientific and philosophical reflections on matter and information, as presented above, may elucidate cosmological claims inherent in religious traditions, here Christianity. At the same time I hope to give evidence of the extent to which a premodern text, such as the Prologue to the Gospel of John, remains sensitive to different aspects of the concept of information, as laid out above. Thus the idea of the divine Logos in the Gospel of John relates to both cutting and shaping information (Logos as "Pattern"), to the life-informing aspects (Logos as "Life"), and to meaning information (Logos as "Word"). And despite its divine origin, Logos is even perceived as embodied (becoming "Flesh").

The early Patristic exegesis of the first verse of the Gospel of John ("In the Beginning was Logos") shows that Christianity was not from the beginning coined in purely Platonic terms, but rather evidences an influence from Stoic physics, the standard philosophy in the Roman empire between 100 BCE and 150 CE. Unlike Platonism, Stoic physics was basically materialist, while giving ample room for the "informational" (and even rational) aspects of the material world. Earlier New Testament scholarship, however, has widely neglected the Stoic influence on Early Christian thinking. Nineteenth-century historian and liberal theologian Adolf von Harnack found in John only a Platonic–Hellenistic thinking. In the early twentieth century existentialist Rudolf Bultmann perceived John as a reflex of a "Gnostic redeemer myth."[12] Moreover, in contemporary Johannine scholarship, especially of Roman Catholic origin, one can notice a tendency to underline the Jewish and Biblical resources behind the Gospel, even to the point of neglecting the cosmological horizon of John.[13] Strands of more recent New Testament scholarship, however, again emphasize the influence of contemporary Stoic physics for both Paul and John, thereby opening up new possibilities for understanding the way in which the theologically reflective writings of Paul and John presuppose a synthesis of Stoic and Middle Platonic ideas, common with many of their contemporaries.[14]

---

[12] See Adolf von Harnack (1892) and Bultmann (1971).

[13] See Brown (1966, p. 524): "In sum, it seems that the Prologue's description of the Word is far closer to biblical and Jewish strains of thought than it is to anything purely Hellenistic". Also R. Schnackenburg (1979, p. 209): "*VV. 1–3 sind keine für sich stehende kosmologische Betrachtung, sondern die erste Strophe eines christlichen Preisliedes auf den Erlöser. So erklärt sich die Bestimmtheit, mit der vom "Wort" personale Aussagen gemacht werden.*"

[14] An important part of this new scholarship is developed within the Copenhagen School of New Testament scholarship; see Engberg-Pedersen (2000). As Gitte Buch-Hansen portrays John:

Hence God and matter are not simply divided into two separate realms, as in Platonism and in Gnosticism; nor is the meaning of the Gospel explained in purely personalist terms. According to this new perspective, early Christian thinking may have more in common with contemporary scientific questions than so-called "modern" existentialist interpretations of Christianity, which presuppose that both God and humanity are divorced from nature.

The Prologue to the Gospel of John (1:1–14) starts out by placing the significance of the historical figure of Jesus in a cosmic perspective. The divine Logos is seen as both the creative and the formative principle of the universe "in the beginning" (John 1:1–5), and as the revealer for all human beings since the dawn of humanity (John 1:9). It is this universally active divine Logos that became "flesh" in the life history of Jesus of Nazareth (1:14). Logos and "flesh" (*sarx*) are thus two correlative concepts, both of cosmic extension.

The divine Logos was said to be "in the beginning" (*en archē*). The Greek term *archē*, like its Latin equivalent *principium*, denotes not only a temporal beginning but also an ontological beginning. The *archē* thus signifies what we today might call ultimate reality: Logos is the (everlasting) Beginning from which all other (temporal) beginnings take place. Logos was therefore also "in God" (*en theō*). Being "in the Beginning" and being "in God"

"Neither Platonic philosophy with its world of shadows and unstable, unreal phenomena, nor apocalypticism with its judgement of this evil world captures the affirmative aspect of God's relation to the world in the Fourth Gospel" (Buch-Hansen, 2007, p. 5). I thank Gitte Buch-Hansen for extensive conversations on these issues, and for constructive input on this section of the chapter.

are correlates, insofar as God is the generative matrix of all that was, is, and will be. Logos, however, is not said to be identical with God (later in the gospel identified as "the Father"). Logos *is* God in the predicative sense of being divine (*theos*), but Logos is not said to be God in the substantive sense (which would have been rendered as *ho theos*, with a definite article).[15] Logos belongs to God, and is God, but is not flatly the same as the reality of God.

Now the Greek term *Logos* can be translated differently. Still today it is most often translated as "Word" in continuation of the Latin translation *Vulgata* dating from the fourth century. Writing around 200 CE, however, the Church Father Tertullian does not even consider this translation. When discussing the meanings of the Greek term *Logos*, he rather points out that in Latin it could be rendered both as *ratio* (rationality) and as *sermo* (speech). Tertullian, however, finds it inconceivable that God should be thought of as "speaking" from eternity, before the temporal beginning. Rather, in God's eternity, Logos must denote the divine rationality or mind (*ratio*), which may involve an inner dialogue (*sermo*), but does not speak out until the creation of the world when there are creatures to speak to. Before creation Logos is not talkative.[16]

Tertullian is here presupposing a distinction from Stoic school philosophy between the "logos inherent in God" (*logos endiáthetos*) and the "outgoing divine logos" (*logos prophórikos*). Among the Greek Church Fathers such as

---

[15] C. K. Barrett: "*theos*, without the article, is predicative and describes the nature of the Word" (Barrett, 1972, p. 130).

[16] Tertullian, *Adversus Praxeas* 5: "*non sermonalis a principio, sed rationalis Deus etiam ante principium.*"

Theophilus of Antioch (c. 190 CE), this Stoic distinction is explicitly used,[17] an indication of the fact that Stoic physics was well known among Christian writers in the Roman Empire.

These early interpretations of John testify that Stoicism (and not only Platonism) has been inspirational for Christian thinking, also among the orthodox, anti-Gnostic Church Fathers. As a consequence, there is the strongest possible link between God's "inner" nature and God's "external" creativity. This interpretation is founded in the very text of the Prologue. For it is about the divine Logos that it is said: "Through this Logos all things came into being, and apart from this nothing came into being" (John 1:3). Applied to the discussions above, one might say that Logos is identified as the *divine informational resource*, which is creative by setting distinctions into the world (carving out 'this' and 'that') and by bringing informational patterns into motion and resonance (combining 'this' and 'that'). We may speak of the Logos as the informational matrix for the concrete forms that have emerged and will emerge in the world of creation.

In Stoic thinking there is no gulf between God and world as in the Platonic tradition, for Logos is all-pervasive as the structuring principle of the universe. Logos thus expresses itself both in the harmonious order of the universe, in the desires of living beings, and in the rational capacities of human beings. Accordingly the Prologue of John states: "In Logos was Life, and that Life was the light of human beings" (John 1:4). "Life" indicates

---

[17] Theophilus of Antioch, *Ad Autolycum* II, 10. I have discussed these texts in more detail in Gregersen (1999).

biological life, especially with respect to the flourishing of life, and by "enlightened human beings" are meant not only specific religious groups such as Christians, but any human being born into the world: "The true Light gives light to any human being who is entering into the world" (John 1:9).[18]

While seeing the divine Logos as the active principle, and the logical structures of the universe as passive qualities, the Stoics retained a distinction between world and God. They claimed, however, that God was no less material than the physical cosmos, just finer and more airy and fiery. Their concept of the material, however, was not the corpuscular theory of the rival school of the Epicureans; rather, they presupposed the cosmos to constitute one homogenous field of energy and matter (related to the elements of Fire–Air and Water–Earth, respectively). "The only way fairly to describe Stoic physical theory would seem to be as a field theory, as opposed to the corpuscular theory of the atomists," says Johnny Christensen in his *Essay on the Unity of Stoic Philosophy*. But the role of the divine Logos is exactly to explain the *unity of differentiation and structure* within cosmos: "'Motion' is most closely connected with structure (*lógos*). It refers to the parts of nature, implying maximum attention to structure and differentiation" (Christensen, 1962, pp. 24, 30).

---

[18] Very often this verse 9 is translated differently, as if it meant the Logos that was about to come into the world. But "coming" (Greek: *erchomenon*) is better attached to "the human being" (*anthrópon*), both in the accusative, than to the Light, which is the subject of the sentence, hence in the substantive position. Apart from the linguistic oddity of the standard translation, it also succumbs to an impatient exegesis, as the incarnation in Jesus does not appear until v. 14.

I do not claim that the Johannine concept of Logos is derived exclusively from Stoic resources, for the Logos concept is semantically flexible, and has Jewish, Stoic and Middle Platonic connotations. However, I do believe that it would be wrong to tag the Gospel of John exclusively on Platonism, or a version of Gnosticism. There is in the Gospel of John no split (Plato: *chōrismos*) between the eternal Logos of God and the Logos at work in creation *within* the one field of physical differentiations, biological life, and human enlightenment. "He was in the world, but the world, though it owed its being to him, did not recognize him. He came to his own, and his own people would not accept him" (John 1:10–11). What is said here eventually is that the Logos is "at home in the universe." The problem is not that there should be principled distance between God and the world; rather the problem lies in the failing human awareness thereof.

Needless to say, Christians departed from the Stoics in their insistence on the pre-material status of the divine Logos. There was a fleshless Logos (*logos asarkos*) before the Logos became incarnate (*logos ensarkos*). It is here that the Christian retained a Jewish and Platonic sense of God's transcendence, while balancing this "Platonizing" element with a strong "Stoicizing" doctrine of the incarnation of the Logos: "Logos became flesh (*sarx*)" (John 1:14).

Like the idea of Logos, the term "flesh" is also semantically flexible, which may well be intentional. "Flesh" can mean simply "body and flesh," thus referring to the historical person of Jesus. This is the case, beyond doubt. But *sarx* can also mean "sinful flesh," thereby intimating that the incarnation of Logos as Jesus Christ already anticipates the death of Jesus for all humankind. What is born by flesh can only give birth to flesh, whereas it is spirit that makes alive (John 3:6). Moreover, the final word of Jesus according to the Gospel of John (19:30) is rendered as follows: "It is accomplished!" The process of incarnation was not fulfilled until the end of the life of Jesus on the cross. But third and finally, flesh usually simply denotes "materiality" in its most general extension, although perhaps especially with respect to its frailty and transitoriness. In this case, one could speak of John 1:14 as involving a notion of *deep incarnation*: The incarnational move of the divine Logos is not just into a particular human person in isolation, "the blood and flesh" of Jesus. The incarnation also extends into Jesus as an exemplar of humanity, and as an instantiation of the "frail flesh" of biological creatures. With the cosmic background of the Prologue in mind, we can now also say that divine Logos, in the incarnation, unites itself with the very basic physical stuff.[19] In other words, the flesh that is assumed in Jesus of Nazareth is not only the boy from Jerusalem, but also a human being, an animal, and the material stuff itself. In Biblical language, God became a human being, a sparrow that flies and falls to the ground, became the green and withering grass, in order to become one with the earthly matter (*sarx*).

This latter interpretation has significant consequences for understanding the relation between God and the material world at large. The divine Logos is then not only locally present in the particular body of Jesus. Logos is present – as Creator and as Redeemer – at the very core

---

[19] The concept of deep incarnation I have earlier developed in Gregersen (2001).

of material existence. The death of Jesus, then, fulfills the self-divesting nature of the divine Logos for all sentient and suffering beings, human or animal. And Logos would be the Light not only for every human being entering the world, but also the "Light of the world" and the "Light of the life" (John 8:12).

Such universalistic interpretation of the Gospel is possible already with the Jewish background of the idea of God having his "home" (*shechinah*) in the midst of the world. But the interpretation is made even more plausible by understanding the Prologue as having a Stoic inspiration. For here Logos is affirmed as the living bond between the ultimate reality of God and the penultimate reality of the world.[20]

Seen from this historical context and applied to today's context of an informational universe, the divine Logos

[20] This interpretation is only warranted if justice is done to the Johannine view that by virtue of this cosmic background, human beings (who are rational beings) are called to re-enact the Truth and the Way in practical discernments. Thus the presence of Logos is not only a theoretical view "to be believed", but "the Truth has to be lived" (John 3:21). In order to make this possible, Jesus Christ is withdrawing from his disciples, in the form of personified Logos. It is *necessary* that Jesus leaves his disciples, it is said, for otherwise the divine Spirit is without place to energize the disciples, to the end that the Spirit "will guide you into all the truth" (John 16:13). "All truth" here also means "doing the right things in practical life"; that is, applying the truth to particular circumstances. Buch-Hansen (2007) even argues that the Spirit is the intentional agent and protagonist of the Jesus story, so that the incarnate Logos, Jesus Christ, is prompted by the life-giving Spirit. As argued by her, also the Johannine idea of Spirit/Wind (*pneuma*) includes a physical dimension, as in John 20:22, when the resurrected Jesus addresses the disciples: "When he had said this, he *breathed on* them and said to them, 'Receive the Holy Spirit'" [my emphasis].

could be seen as the informational resource active in the world of creation, both by generating distinctiveness from within the core of stochastic quantum processes, by channeling energetic drives via thermodynamical processes, by building up and reshaping biological structures, and by facilitating connections and communication at whatever level possible. Some of these aspects have a rather strong law-like character (at least in terms of overall statistics), while others rely on more contingent historical processes.

On this background it would be possible to see a deep congeniality between a Logos Christology explicated in its cosmic framework and contemporary concepts of matter and information. The "flesh" of the material world is by John seen as saturated by the presence of the divine Logos, who has united itself with the world of creation – by creating differences ("cutting information"), by shaping and reshaping ("instructing and building up"), by creating constructive resonances between organisms and their environments ("absorbing and connecting"), and by making meaning and communication possible ("making sense" of things).

On such a background it is perhaps also possible to affirm the beautiful *Hymn to Matter*, formulated by the Jesuit paleoanthropologist Teilhard de Chardin in a distinctive moment of amazement after having experienced the shock of World War I. Nonetheless, Teilhard was willing to accept the world as it is, including its harshness and riskiness. While one might not want to share Teilhard's overall evolutionary progressivism, his hymn to matter shows a deep sense of the unity between a creative God and a material world saturated by a divine presence:

Blessed be you, harsh matter, barren soil, stubborn rock: you who yield only to violence, you who force us to work if we would eat.

Blessed be you, perilous matter, violent sea, untamable passion: you who unless we fetter you will devour us.

Blessed be you, mighty matter, irresistible march of evolution, reality ever new-born; you who, by constantly shattering our mental categories, force us to go ever further and further in our pursuit of the truth.

Blessed be you, universal matter, immeasurable time, boundless ether, triple abyss of stars and atoms and generations: you who by overflowing and dissolving our narrow standards or measurement reveal to us the dimensions of God...

I acclaim you as the divine milieu, charged with creative power, as the ocean stirred by the Spirit, as the clay moulded and infused with life by the incarnate Word.

(Teilhard, 1978, pp. 75–76)

References

Barrett, C. K. (1972). *The Gospel According to St. John.* London: SPCK.

Brown, R. E. (1966). *The Gospel According to John (i–xii).* New York: Doubleday.

Buch-Hansen, G. (2007). *It is the Spirit that Makes Alive (6:63). A Stoic Understanding of pneuma in John.* Copenhagen: Copenhagen University.

Bultmann, R. K. (1971). *The Gospel of John: A Commentary.* Philadelphia, PA: Westminster Press.

Christensen, J. (1962). *An Essay on the Unity of Stoic Philosophy.* Copenhagen: Munksgaard.

Davies, P., and Gribbin, J. (1992). *The Matter Myth: Dramatic Discoveries that Challenge our Understanding of Physical Reality*. New York: Simon & Schuster.

Deacon, T. W. (2003). The hierarchic logic of emergence: Untangling the interdependence of evolution and self-organization. In *Evolution and Learning: The Baldwin Effect Reconsidered*, eds. B. H. Weber and D. J. Depew. Cambridge, MA: MIT Press, 273–308.

de La Mettrie, J. O. (1748). *Machine Man and Other Writings*, ed. Ann Thomson (1996). Cambridge: Cambridge University Press.

Einstein, A., and Infeld, L. (1938). *The Evolution of Physics*. New York: Simon & Schuster.

Emmeche, C. (1999). The Sarkar challenge to biosemiotics: Is there any information in the cell? *Semiotica*, 127(1/4): 273–293.

Engberg-Pedersen, T. (2000). *Paul and the Stoics*. Louisville, KY: Westminster John Knox Press.

Flores, F. (2005). The Equivalence of Mass and Energy, *Stanford Encyclopedia of Philosophy*, originally accessed 2 August 2006; substantive revision 2007 (http://plato.stanford.edu/entries/equivME).

Gregersen, N. H. (1998). The idea of creation and theory of autopoietic processes. *Zygon: Journal of Science & Religion*, 33(3): 333–367.

Gregersen, N. H. (1999). I begyndelsen var mønsteret. *Kritisk Forum for Praktisk Teologi*, 75: 34–47.

Gregersen, N. H. (2001). The cross of Christ in an evolutionary world. *Dialog: A Journal of Theology*, 40(3): 192–207.

Gregersen, N. H. (2007). Reduction and emergence in artificial life: A theological appropriation. In *Emergence from Physics to Theology*, eds. N. Murphy and W. Stoeger. New York: Oxford University Press, 284–314.

Hanson, N. R. (1962). The dematerialization of matter. *Philosophy of Science*, 73(1): 27–38.

Hoffmeyer, J. (2008). *Biosemiotics. An Examination into the Signs of Life and the Life of Signs*. Scranton and London: University of Scranton Press.

Laplace, P. S. (1813). *The System of the World*, vols 1–2, trans. J. Pond (2007). Whitefish, MT: Kessinger Publishing.

Lloyd, S. (2006). *Programming the Universe*. New York: Knopf.

Mayer, J. R. (1842). Bemerkungen über die Kräfte der unbelebten Natur. *Philosophisches Magazin*, 24: 371–377.

Mayer, J. R. (1980). Remarks on the forces of inorganic nature. In *Darwin to Einstein: Primary Sources on Science & Belief*, eds. N. G. Coley and V. M. D. Hall. Harlow, UK: Longman, 68–73.

Newton, I. (1952). *Opticks, or A Treatise of the Reflections, Refractions, Inflections & Colours of Light*, 4th ed (1730). New York: Dover Publications.

Peacocke, A. (2007). *All That Is: A Naturalistic Faith for the 21st Century*, ed. P. Clayton. Minneapolis: Fortress.

Puddefoot, J. C. (1996). Information theory, biology, and christology. In *Religion and Science: History, Method, Dialogue*, eds. M. Richardson and W. J. Wildman. New York: Routledge, 301–320.

Richards, R. J. (1987). *Darwin and the Emergence of Evolutionary Theories of Mind and Behavior*. Chicago, IL: University of Chicago Press.

Russell, B. (1961). Introduction to *A History of Materialism*, by F. A. Lange (1925). In *The Basic Writings of Bertrand Russell 1903–1959*, eds. R. Egner and L. E. Denonn (1961). New York: Simon & Schuster, 237–245.

Schnackenburg, R. (1979). *Das Johannesevangelium. Einleitung Kommentar Teil 1*. Freiburg: Herder Verlag.

Stubenberg, L. (2005). Neutral monism, *Stanford Encyclopedia of Philosophy*, originally accessed 31 July 2006; substantive revision 2010 (http://plato.stanford.edu/entries/neutral-monism).

Teilhard de Chardin (1978). *The Heart of Matter*. London: Collins.

Tertullianus (c. 200). *Adversus Praxeas*. In *Patrologia Latina*, vol. 2, ed. J. P. Migne (1857). Turnhout: Brepols, 160.

Theophilus of Antioch (c. 190) *Ad Autolycum*. In *Patrologia Graeca*, vol. 6, ed. J. P. Migne (1866), Turnhout: Brepols, 1064.

Thomson, W. (1980). On the dynamic theory of heat. In *Darwin to Einstein*, eds. N. G. Coley and V. Hall. Harlow, UK: Longman, 84–86.

von Harnack, A. (1892). Über das Verhältniss des Prologs des vierten Evangeliums zum ganzen Werk. *Zeitschrift für Theologie und Kirche*, 2(3): 189–231.

Weaver, W. (1949). Recent contributions to the mathematical theory of communication. In *The Mathematical Theory of Communication*, eds. C. E. Shannon and W. Weaver. Urbana: The University of Illinois Press, 94–117.

Wolfram, S. (2002). *A New Kind of Science*. Champaign, IL: Wolfram Media.

Zemann, J. (1990). Energie. In *Europäische Enzyklopädie zu Philosophie und Wissenschaften*, ed. H. J. Sandkühler et al. Hamburg: Felix Meiner, 694–696.

16

What is the 'spiritual body'?: on what may be regarded as 'ultimate' in the interrelation between God, matter, and information

MICHAEL WELKER

~

This chapter operates from a theological perspective – broadened, however, by information on the development of classical philosophy and metaphysics and some experience in the global science-and-theology discourse of the last 20 years. I ask the question: Can we imagine and penetrate the reality classical theology had in mind when it spoke of the 'spiritual body'? And beyond that, can we convince non-theological mindsets that this concept not only makes sense in the orbit of religion, but that it has illuminating power beyond this realm because it is firmly rooted in a reality, and not just confined to one complex mode of discourse?

The preparation for this task requires a few sophisticated preliminary steps. *First* we have to differentiate 'old-style' and 'new-style' metaphysics as two possible frameworks for the approach. *Second*, we have to discern an understanding of creation in the light of Biblical

Information and the Nature of Reality: From Physics to Metaphysics, eds. Paul Davies and Niels Henrik Gregersen. Published by Cambridge University Press © P. Davies and N. Gregersen 2010, 2014.

creation accounts and in the light of 'old-style' meta-physics. *Third*, on the basis of the Biblical creation accounts, we see that the notion of a creator as a mere sustainer of the universe is spiritually not satisfying and salvific. *Fourth*, this will prepare us for an understanding of the role of the resurrection in divine creativity in general, and provide an understanding of the nature and the importance of the 'spiritual body' of Jesus Christ in particular. *Fifth*, we will try to comprehend the transformative power of this spiritual body and the involvement of human beings and other creatures in it. On this basis we want to engage non-theological academic thinkers by asking them whether the sustaining, rescuing, and ennobling interaction between God, creation, and spiritual information can find analogies in their realms of experience, and whether it can challenge reductionistic concepts of matter. This question will be guided by a metaphysical approach in the 'new style'.

16.1 Differentiating 'old-style' and 'new-style' metaphysics

My proposal to differentiate between 'old-style' and 'new-style' metaphysics does not imply that 'old-style' metaphysics is outdated and should be simply replaced by 'new-style' metaphysics. 'New-style' metaphysics is a constructive reaction to the lament that after Kant, metaphysics as the production of ultimate and closing thoughts about total reality is no longer possible. To be sure, philosophy has to face the dilemma that in late modern societies, in their academic as well as in their religious settings, a plurality of forms of life and rationalities has established itself, and can no longer be convincingly ordered in 'a hierarchy of the more or less valuable.'[1] In this situation, 'old-style' metaphysics can be engaged in order to relativize the current epistemic setting at least in the academy in the West – to relativize it with world views and cognitive claims from past epochs. In a more modest and empirically argumentative way, 'new-style' metaphysics responds to this challenge by reducing the metaphysical claim to an exploration of two areas of discovery and research.

It was Alfred North Whitehead who in a most helpful way differentiated between old- and new-style metaphysics (without using these terms). On the one hand, he states in terms of 'old-style' metaphysics: 'By "metaphysics" I mean the science which seeks to discover the general ideas which are indispensably relevant to the analysis of everything that happens' (Whitehead, 1960, p. 82). On the other hand, in the mode of 'new-style' metaphysics, he speaks not of 'the,' but of 'a' metaphysics and of a 'metaphysical description' that 'takes its origin from one select field of interest. It receives its confirmation by establishing itself as adequate and as exemplified in other fields of interest' (Whitehead, 1960, p. 86f). Whitehead makes it clear that 'a metaphysics' can emerge from different homelands: mathematics, a science, religion, common sense. As soon as a solid bridge theory can be established between at least two areas of interest by indicating that basic ideas, concepts, and intellectual operations can work in both areas, we are in the process of a 'metaphysical description' and on the way to 'a metaphysics' – which I call 'new-style' metaphysics.

---

[1] As a representative voice, see Habermas (1987, p. 434).

446

It is the 'bottom-up' approach against the 'top-down' thinking that is characteristic of a 'new-style' metaphysics. Such a metaphysics tries to cultivate common sense, to challenge it, and move it to higher levels of insight by confronting it with specific 'fields of interest' that require specific modes of thinking in order to be adequately explored. The differences between common-sense thought and the thinking required to access at least one of these fields of interest, but also the differences between the cognitive explorations of at least two of these fields of interest (for example in science and theology), provide the impulses to develop a 'new-style' metaphysics.

16.2 Creation according to 'old-style' metaphysics and according to classical Biblical accounts

Most theological and philosophical thinking about 'creation' has been dominated by the concepts of 'bringing forth' and 'dependence'. Creation as *creatura* was nature, the cosmos, or vaguely conceived totality as brought forth and dependent on one or several transcendent power(s) or will(s) or personal entity(ies), mostly named God or gods. Creation as *creatio* was the activity or energy of bringing forth, keeping in dependence or even in 'ultimate' dependence and at the same time sustaining nature, the cosmos, the totality (sometimes explicitly also referring to culture and history). Connected with this type of thinking, ideas and concepts of God such as 'the all-determining reality' (Bultmann, Pannenberg), the 'ground of being' (Tillich), the 'ultimate point of reference' (G. Kaufman), the 'whither of absolute dependence' (Schleiermacher) were highly en vogue.[2]

In striking contrast, the classical Biblical Priestly creation account in Genesis 1 offers a much more subtle picture. Through the word of God, chaotic matter becomes enabled not only to win forms and shapes, energy, and life. The heavens, the stars, the earth, the waters, and the humans are to actively participate in God's creative energy and power. The same verbs used for the divine process of creating are also used for the co-creativity of what and who are created. Against a widespread fear of a 'synergistic' confusion of God and creature in the case of co-creativity, it has to be recognized that the Biblical creation account does not think in one-to-one structures (God and creation, God and world, God and the human being). The account thinks in one-to-many structures in which selected creatures gain a graded share in the creative divine activity. In various ways selected creatures participate in the formation of creation. The heavens part, the stars govern the times and festive days, the earth brings forth creatures, and the humans are assigned the task to rule over creation and thereby reflect the image of God.

In this one-to-many relationship, no creature has the power to act in God's stead, yet the power of the co-creative creatures is sizable. This power not only enables humans as well as other creatures to exercise their creaturely freedom and act independently; it also makes creaturely self-endangerment and self-destruction possible. There are several indications in the creation account that support this realistic reading against all metaphysical

---

[2] See Welker (1999a), especially Chapter 1.

'perfect-watchmaker' illusions (which, as a rule, come with the theodicy question in the backpack). The co-creative creatures remain creatures. Despite their powers, they are no gods (as other ancient creation accounts would have it). Neither the heavens nor the sun, moon, and stars, nor the earth are divine powers. And the 'slave-holder and conqueror' language in the infamous 'call to dominion' indicates that a constant conflict between the humans and the animals in a common area of nourishment has to be regulated. The humans clearly become privileged, yet in all their self-interest and privileging of their own self-reproduction and spreading over the earth they have to mirror the image of God to the other creatures, and that means in ancient royal imagery: the exercise of justice and mercy towards the weak.

Whoever does not like this sober picture of creation that takes seriously the fact that life is finite, that it lives at the expense of other life, and that the co-creative power of the creatures brings the risks of self-endangerment, seems to find an argument against this understanding in the Priestly account itself, an argument impossible to beat. Doesn't, after all, the Biblical creation account press us to embrace the speculative metaphysical view by the repeated assertion of the 'goodness' of creation? Repeatedly, the co-creative creaturely entities and the areas of creation are seen by God as 'good' (Genesis 1:4, 10, 12, 18). The works of creation viewed all together are even judged to be 'very good' (Genesis 1:31): 'God saw everything that He had made and lo, it was very good.' 'Good' (*tob* in Hebrew) means 'conducive to life.' But a creaturely and even co-creative existence, even if it is highly conducive to life, has not yet reached the level of divine glory. The difference

449

between God's glory and the creation judged by God to be 'good' and 'very good' is still maintained. Creation is not paradise.

16.3 'Creation' itself points beyond God as a mere sustainer of the universe

It belongs to the rituals in the science-and-theology or science-and-religion discourses and their public radiations that some of the scientists connect their summarizing perspectives with religious awe and theological respect for the power and wisdom of a divine creator. The power of mathematics and rational thinking in illuminating hidden secrets of nature, observations of unquestionable beauty and astounding rhythmic orders, the fecundity of life and its potential to generate 'higher forms' are named in order to support such views.[3] Others, however, leave such discourses with the summary: 'The more I looked at the universe, the more I found it pointless.' Or they think along Whitehead's words: 'Life is robbery and requires justification.'

At this point, we become aware of the fact that any 'natural awareness of the divine' is connected with pressing problems – as Calvin showed so powerfully at the beginning of his *Institutes of the Christian Religion* in 1559. Any perspective on God as creator and sustainer of the world will never overcome the ambivalence and ultimate inconsolability of a 'natural' theology of creation. Calvin calls this 'sense of Deity' – which he sees beyond dispute – 'fleeting and vain' (Calvin, 1559, I, 3, 3).

[3] See Polkinghorne and Welker (2001).

If we do not overrule the realistic experience of creation with 'old-style' metaphysics, we have to acknowledge that a power merely sustaining the universe – impressive as it is – is ultimately not worth being called 'divine'. And the instantiation, the 'whither' of this power, is not worth being called 'God'. Confronted not only with the finitude of life, but also with the fact that life can live only at the expense of other life, and that the co-creative power of the creatures is full of self-endangerment and destructive potentials, we have to ask for the saving and ennobling workings of the creative God in order to overcome the deep ambivalence just depicted.

'Saving and rescuing' can in this case, as we easily see, not just mean repairs in the course of natural processes. To be sure, our life and world experiences of birth and healing, of forgiveness, reconciliation, and peace mirror the depth of God's creative care and guidance. And these experiences can cause gratitude and joy, praise and glorification. But the haunting question remains of whether God's creativity can ultimately overcome the finitude and deep ambivalence of creaturely life itself. This question cannot be raised and answered without addressing the difficult topics of eschatology: new creation and the resurrection.

16.4 Divine creativity in the resurrection and spiritual body of Christ

The resurrected Jesus Christ is not the resuscitated pre-Easter Jesus of Nazareth. Although a few witnesses of the resurrection in Luke – and certainly all sorts of Christian fundamentalists and their critics – seem to confuse resurrection with resuscitation, the Biblical insights on this topic are clear. The Biblical texts report experiences of homage (*proskynesis*) in which witnesses fall down in worship in the face of a theophany, although this revelation of God is mixed with doubt at the same time.[4] Jesus' resurrection is a reality that, on the one hand, exhibits characteristics of something sensory, while on the other hand retaining the character of an appearance, even an apparition. The Emmaus story is especially revealing: the eyes of the disciples are kept from recognizing the resurrected one. By the ritual of the breaking of the bread, their eyes are opened. Then in the next verse it says: 'And he vanished out of their sight.' Instead of complaining about a spooky event, about just having seen a ghost, the disciples remember a second experience that had not seemed to be a revelation at first: 'Did not our hearts burn within us while he talked to us on the road, while he opened to us the scriptures?' (Luke 24:30ff).

The witnesses recognize the resurrected one not only by his salutation, the breaking of the bread, the greeting of peace, the way he opens up the meaning of Scripture, and other signs, but also through his appearances in light. These clearly speak against any confusion of resurrection with physical resuscitation. A multitude of diverse experiences of encounter with Christ brings about certainty that Christ is, remains, and will be 'bodily' present among us. By contrast, the stories of the empty tomb show that a single moment of revelation alone, even if it is a spectacular one from heavenly messengers, is not in itself enough to

---

[4] See also Eckstein and Welker (2006) and Peters, Russell, and Welker (2002, especially pp. 31–85).

cause belief. Instead, what remains after the empty grave are fear, amazement, and silence (Mark). Meanwhile, the belief that the corpse has been stolen is disseminated, used for propaganda purposes, and becomes widespread (John and Matthew). According to Luke, the visions at the empty tomb are dismissed as 'women's chatter'.

The certainty that Christ is risen does not signify, however, that he is present in the way that the pre-Easter Jesus was. In fact, the complete fullness of his person and his life is now present 'in Spirit and in faith'. This presence 'in Spirit and faith' is hard to comprehend, not only to naturalistic and scientistic thought, which tends to fixate instead continually on the pros and cons of physical resuscitation. By contrast, the fullness of the person and life of Christ accentuates the community of witnesses in Spirit, faith, and canonical memory. In this way, the entirety of Jesus' life, his charisma and his power, is present and efficacious in the resurrected and exulted one.

The presence of the resurrected one conveys the powers of love, forgiveness, healing, and his passion for children, the weak, the rejected, the sick, and those in misery. Further, the power to confront the so-called 'powers and principalities' begins to take shape in his presence: for example in conflict with political and religious institutions in the search for truth and demands for real justice. The person and life of Jesus Christ unleashes normative and cultural renewals and other creative impulses. The presence of the resurrected Christ is realized among the witnesses through many signs – including small ones – of love, healing, forgiveness, devotion, acceptance, and the passionate search for justice and truth. In this often inconspicuous way, Christ and the kingdom of God are 'coming'.

Besides this emergent coming for which Christians pray in the Lord's Prayer, the Biblical traditions also offer visions of the final coming of the Son of Man. They deal with eschatological visions, and these are necessarily visions because the resurrected and exulted one will not come only in a specific year or to a specific area of the world. The resurrected and exulted one comes in all times and to all areas of the world. He judges, as the Apostle's Creed says, 'the living and the dead'. This is a vision that necessarily transcends all merely natural and empirical conceptions. But it is this important and healing vision that opposes all explicit and implicit egoisms of particular cultures and eras.

If we only have a vision of the Son of Man coming from the heavens with his angels, 'we are of all people most to be pitied', to echo Paul in 1 Corinthians 15. The talk about the 'coming' Christ becomes comforting because the one who is 'coming' will not be revealed for the first time only at the end of all times and eras, but rather, he is already among us now as the crucified and resurrected one; and because the crucified and resurrected one is the one who, in the historical pre-Easter Jesus, has been revealed in his incarnational nearness of his human life and work. For this reason, we cannot separate the memory of the historical Jesus from the realization that the crucified and resurrected one is present and will 'come again' for his full parousia. The creating and saving God is present here, surrounding and carrying his creatures in his 'Yes' to life against the powers of sin and death. These powers are dramatically depicted at the cross of Christ.

At the cross of Christ, Jesus is condemned in the name of politics and in the name of religion. He is executed in the name of both Jewish and Roman law. Even public opinion is against him: 'Then they cried out again, "Crucify him!"' (Mark 15:13f. par.). Jews and Gentiles, Jews and Romans, natives and foreigners all agree. All principalities and powers work together, and the worldly 'immune-systems' collapse. The reciprocal checks and balances between religion, politics, law, and morality fail in the event of the cross. Conflicts between the occupiers and the occupied, the world superpower and an oppressed people, are simply glossed over. Even the disciples betray Jesus, abandoning him and fleeing, as the tradition of the Last Supper, the Gethsemane story, and the 'night of betrayal' make all too clear.

The cross reveals, as the Biblical texts say, 'the world under the power of sin'. It reveals 'the night of godforsakenness', not only of Jesus himself – but of the whole world. The cross reveals the presence of this dire need and misery, not only in Jesus' hour of death, but as a real and present danger in all times. The resurrection liberates from this night of godforsakenness. God's activity alone, and not human initiative, brings salvation. The true saving power and vital necessity of the resurrection only become manifest against the background of the cross. That God, and God alone, brings salvation to humanity becomes recognizable in view of the harrowing possibility and reality that, despite the best intentions and the best systems, humanity alone is doomed. Even the 'good law' of God can become fully corrupt and be abused by humanity under the power of sin. Perversions of religion, law, politics, and public opinion then triumph. Therefore it is crucial to recognize that God has saved and saves humanity, which is completely lost without God. How God saves is also crucial: in a powerful yet emergent way without loud fanfare or drumbeats. As impressive as the Isenheim Altarpiece's portrayal of the resurrection may be, the witnesses of the resurrection in the Biblical traditions describe the reality of God's salvific work in quite a different way.

16.5 The transformative power of the spiritual body: sustenance, rescue, and ennoblement into eternal life

Although the experiences of the resurrected Christ both of the first witnesses and of contemporary witnesses do have the character of visions, memories, and anticipatory imaginations, they are not mere mental or psychic phenomena. They respond to the self-presentation of the resurrected and elevated Christ in his post-Easter body and they participate in his real life. The structured pluralism of the canonic witnesses, the structured pluralism of the ecumenical witnesses, the structured pluralism[5] of a multidisciplinary theology, and the polyphony of the individual witnesses in truth-seeking communities work time and again against illusionary productions of Jesus-images, wishful Jesus-morals, and Jesus-ideologies. It is the faithful realistic response to his presence and his word that critiques and purifies the witness to Christ and to the workings of the Triune God, and thus saves it from being confused with all sorts of self-made religiosity.

[5] It is crucial to differentiate clearly a context-sensitive multisystemic pluralism from a mere 'plurality' of individuals, groups, and their various goals and opinions. The first constellation challenges us to understand a complex structure and circulation of power; the second presents just a soft relativism. By comparison, see Welker (1999b, pp. 9–23; 2000a).

In order to do so, it is crucial to respect both continuity and discontinuity between the life and body of the pre-Easter Jesus, of the resurrected and exulted Jesus Christ, and Christ as the ultimate Judge and Saviour of the world in his parousia. The amazing continuity between the pre-Easter and the post-Easter body is described by Paul with the imagery of the seed and the full-grown plant (1 Corinthians 15:36–38, 44). Yet this amazing continuity is correlated with almost frightening discontinuities: 'the *dying* of the seed and an act of [*new*] *creation* by God (15:38, my italics). Our whole perishable person will be transformed (*metaschematizo*, Philippians 3:21) into a new and imperishable heavenly personality that will be qualitatively different from our first. It will be – thank God – much better!' (Lampe, 2002).

Both continuity and discontinuity are expressed in the term 'spiritual body'. As Paul differentiates between 'flesh' (*sarx*, as perishable matter) and 'body' (*soma*, as matter shaped by mind and Spirit into a living spiritual existence bearing information and giving information), he can perceive continuity and discontinuity in the following way: the body as flesh and as dominated by non-divine powers will decay and die; the body as the spiritual body will be recreated by God's grace in the resurrection.[6] Although 'flesh' is definitely doomed to decay and death, it is full of energies and logics of self-sustenance and self-perpetuation. However, as these energies fall short

---

[6] It is most important not to associate 'flesh' with an understanding of matter in a Newtonian sense as 'solid, massy, hard, impenetrable, moveable particles.' The notion of 'flesh' is not to be confused with 'material stuff' without any information.

of aiming at the existence of the 'spiritual body', they are bound to 'sin and death'.

There would be no substantial hope for our lives if there was not a continuity between our bodily existence on earth – undeniably also shaped by the flesh – and our spiritual body, shaped by the powers of faith, love, and hope. Paul challenges the Corinthians who want to connect Christ and the soul here and in eternity, but want to leave room for any behaviour in terms of sex and food, as the earthly body would die in any case. 'In Paul's holistic perspective... the reality of salvation is not *another* reality apart from the outer everyday life, not just a religious reality for the *inner* life of a person. It grasps and embraces the whole of human existence, the entire personality... For exactly this reason, Paul talks about "resurrection" and not of such things as "spiritual immortality" and "ascending souls"' (Lampe, 2002, pp. 104f).

Connected with this anthropological realism is an eschatological realism that sees any perspective on creation already in the light of the new creation. This eschatological realism affirms that the creative God is not a mere sustainer of the world, as this world is full of ambiguity and despair because of creaturely co-creative freedom and its potentials to misuse; it is full of ambiguity and despair because of the inert brutality and finitude of life in the flesh. The mere affirmation that the Triune God opens much richer perspectives for creation than its continuation until a timely or untimely death is not strong enough to sustain a viable faith and hope towards an eternal existence in a spiritual body. According to Paul, it is rather the presence of the resurrected Christ – in continuity and discontinuity with his pre-Easter life and body and the rich spiritual orientation and information given with this presence – that opens up a totally new perspective. Those who live in Christ as members of his body – consciously or unconsciously – are transformed into his likeness and are preserved towards the eternal life of God.

But it is not only the life of the believers and followers of Christ that gains a salvific perspective through its participation in this spiritual body. In the celebration of the Holy Communion, the Eucharist, the 'elements' bread and wine participate in the edification of the spiritual body, too. The gifts of creation (not just gifts of nature, but gifts of the interaction between nature and culture and thus already richly blessed by the working of the Holy Spirit!)[7] become gifts of 'new creation.' Bread and wine not only symbolically edify the natural bodies of the community assembled. As 'bread and wine from Heaven', as the body and blood of Christ, they edify the members of the body of Christ, the members of the 'new creation,' the bearers of the fruits and gifts of the Holy Spirit. Here the continuity between creation and new creation, between creation old and new becomes palpably present in the middle of the overwhelming discontinuity. It is the spiritual information – to call it thus in search of a more appropriate term – that operates on the material fleshly bodies and minds through the presence of Christ in word and sacrament, causing sometimes dramatic but most often only very calm emergent transformations. It remains to be discussed whether this process can mirror analogies in non-religious and even scientific areas of exploration and interest. With the help of 'new-style' metaphysics, we should try to explore such

7 See Welker (2000b).

analogies in the interactions between 'God, matter, and information', or rather between 'God, earthly and fleshly creatures, and spiritual orientation'.

16.6 Some lessons and inspirations for interdisciplinary dialogues on the topic

In this respect, the anthropology of Paul can provide some clues, when we look at his description of the activities of the spirit. From both theological as well as anthropological perspectives, the Spirit enables co-presence, contact, and even interaction with those who are absent, respectively a presence in absence. Through his Spirit, the invisible God communicates with the human spirit and imparts to it creative impulses. But the communicative power of the spirit can also be vindicated in an anthropological bottom-up approach. According to Paul, even those who are absent can have authentic contact with others 'in the spirit', despite their different locations in space and time. By remembering his own visits, his teaching and preaching, and through his petitions before God, but also through the letters and messages of others, Paul is present to the community 'in the spirit'. This presence is not merely a figment of his imagination.

Paul sees himself becoming 'spiritually' present in the community. In 1 Corinthians 5, he describes this process of spiritual communication and co-action: 'For though absent in body, I am present in spirit; and as if present I have already pronounced judgement in the name of the Lord Jesus...When you are assembled...my spirit is present with the power of our Lord Jesus' (1 Corinthians 5:3–4). 'The name' and 'the power of our Lord', and

certainly the Spirit of God (though not expressly men-
tioned in this passage), play an important role here in the
connection of the community – even in a very general
way: that is, with and without bodily co-presence. However, one does not need to make reference to the 'Holy
Spirit' in order to understand the spiritual process of communication between Paul and the Corinthians. We can
explain basic functions of the human spirit without direct
reference to theological realms.

Memory and imagination are not just 'mental con-
structs' in the 'inner subjectivities', which in a strange way
happen to connect and intersect, thus allowing common
understanding, consensus, and the guided common search
for truth. As instances, which support individual certainty,
communal consensus, and the oriented progress of truth-
seeking communities,[8] they certainly do have 'points of
reference' in natural space–time and the matter correlated
with it. But in the mediation from empirical experience to
shared forms of memory[9] and sustainable common imagination, there is a 'spiritual loadedness' of the experienced
material reality and a transfiguration of it. In secular terms,
this transformation and the mental participation in it has
been beautifully expressed by Wordsworth in his famous
poem 'Daffodils':

> Ten thousand saw I at a glance,
> Tossing their heads in sprightly dance...
> I gazed – and gazed – but little thought
> What wealth the show to me had brought:
> For oft, when on my couch I lie

[8] The texture of truth-seeking communities is described in the last
chapter of Polkinghorne and Welker (2001).
[9] See Assmann (1992, 2000); Welker (2008, pp. 321–331).

In vacant or in pensive mood,
They flash upon that inward eye
Which is the bliss of solitude;
And then my heart with pleasure fills
And dances with the daffodils.

References

Assmann, J. (1992). *Das kulturelle Gedächtnis.* München: Beck.

Assmann, J. (2000). *Religion und kulturelles Gedächtnis.* München: Beck.

Calvin, J. (1559). *Institutes of the Christian Religion,* trans. Henry Beveridge (2008). Peabody, MA: Hendrickson Publishers.

Eckstein, H.-J., and Welker, M., eds. (2006). *Die Wirklichkeit der Auferstehung,* 3rd ed. Neukirchen-Vluyn: Neukirchener Verlag.

Habermas, J. (1987). Metaphysik nach Kant. In *Theorie der Subjektivität,* eds. K. Cramer et al. Frankfurt: Suhrkamp, 425–443.

Lampe, P. (2002). Paul's concept of a spiritual body. In *Resurrection: Theological and Scientific Assessments,* eds. T. Peters, R. Russell, and M. Welker. Grand Rapids: Eerdmans, 103–114.

Peters, T., Russell, R., and Welker, M., eds. (2002). *Resurrection: Theological and Scientific Assessments.* Grand Rapids: Eerdmans.

Polkinghorne, J., and Welker, M. (2001). *Faith in the Living God. A Dialogue.* Philadelphia: Fortress; London: SPCK.

Welker, M. (1999a). *Creation and Reality.* Philadelphia: Fortress.

Welker, M. (1999b). Was ist Pluralismus? In *Wertepluralismus. Sammelband der Vorträge des Studium Generale der Ruprecht-Karls-Universität Heidelberg im Wintersemester 1998/99.* Heidelberg: C. Winter.

Welker, M. (2000a). *Kirche im Pluralismus*, 2nd ed. Gütersloh: Kaiser.

Welker, M. (2000b). *What Happens in Holy Communion?* Grand Rapids: Eerdmans.

Welker, M. (2008). Kommunikatives, kollektives, kulturelles und kanonisches Gedächtnis. In *Jahrbuch für Biblische Theologie*, Bd. 22: Die Macht der Erinnerung. Neukirchen-Vluyn: Neukirchener Verlag, 321–331.

Whitehead, A. N. (1960). *Religion in the Making*, (originally 1926). New York: Meridian.

INDEX

Index

understanding
as constituent of information,
365
as distinct reality, 363, 366
philosophical discussion of, 72,
219–24, 229, 374
unified theory of physics, 87
universe, *see also* cosmic bit count;
cosmology; multiverse
model; ultimate reality
as clockwork, 4, 94, 123, 225
as code lacking, 168
as holistic system, 359, 368
as information system, 384
as informational, 198, 384, 388,
394, 397, 399–402, 438
as mathematical, 90
as polynomial or exponential,
107
bearing revelatory meaning, 387
billiard-ball, 134
carrying information, 104, 385
compatible with omnipotent
Creator?, 56, 58
computational, 4, 104, 113,
118–32, 198, 423–27
contingency of, 85–86, 412
deterministic, *see* determinism
expanding, 15, 265
expanding or contracting, 33
narratively informed or
mechanically designed?, 398
naturalistic view of, 377
non-ergodicity of the, 243
origin of, 55, 359, 362
quantum, 135
reasons for existence of, 371–73
registering itself, 4, 424
static or contract, 36
structured by Logos, 414, 432,
434
two-stock, 405

vacuum, 32–33, 41, 83, 144, 415
vacuum energy, 38
density of, 38
van de Vijver, G., 74, 76
van Till, H., 327, 355
Varela, F. J., 270, 310
Venter, J. C., 288, 311
von Balthasar, H. U., 53, 75
von Goethe, J. W., 227, 235
von Harnack, A., 430, 442
von Neumann, J., 74, 144, 147,
153, 204, 367, 409
von Weizsäcker, C. F., 68

Wallace, A. R., 159
Ward, K., xiv, 9, 116, 357
Waters, P. J., 241, 259
wave function, 8, 103–10, 113
Weaver, W., 124, 132, 163, 185,
215, 419, 442
Weber, B. H., 237, 258–60, 322,
352, 355, 440
Weinberg, J., 56, 79
Weinberg, S., 39, 63, 79
Weisheipl, J., 20, 45
Weismann, A., 157, 159, 163–66,
185
Weizsäcker, C. F., 68
Welker, M., xiv, 10–11, 444, 462
Wheeler, J. A., 74, 90, 94–96, 116,
367, 369, 379
Whitehead, A. N., 62, 79, 387,
403, 462
as pan-experientialist, 62
life is a robbery, 450
on evil, 401
on matter, 50
on old- and new-style
metaphysics, 445
on pessimism, 396
on structuring principle, 382
ontology of, 62, 148

Printed by Printforce, United Kingdom